Analytical Evaluation of Nonlinear Distortion Effects on Multicarrier Signals

OTHER COMMUNICATIONS BOOKS FROM AUERBACH

Analytical Evaluation of Nonlinear Distortion Effects on Multicarrier Signals

Teresa Araújo
Rui Dinis

CRC Press
Taylor & Francis Group
Boca Raton London New York

CRC Press is an imprint of the
Taylor & Francis Group, an **informa** business

CRC Press
Taylor & Francis Group
6000 Broken Sound Parkway NW, Suite 300
Boca Raton, FL 33487-2742

First issued in paperback 2021

© 2015 by Taylor & Francis Group, LLC
CRC Press is an imprint of Taylor & Francis Group, an Informa business

No claim to original U.S. Government works

ISBN 13: 978-1-138-89441-9 (pbk)
ISBN 13: 978-1-4822-1594-6 (hbk)

Contents

List of Abbreviations

ADC Analog-to-Digital Converter

ADSL Asymmetric Digital Subscriber Line

AWGN Additive White Gaussian Noise

BA Block Assignment

BER Bit Error Rate

BPSK Binary Phase-Shift Keying

BS Base Station

CAS Carrier Assignment Scheme

CIR Channel Impulse Response

CP Cyclic Prefix

DAB Digital Audio Broadcasting

DFT Discrete Fourier Transform

DMT Discrete Multitone

DSL Digital Subscriber Line

DVB Digital Video Broadcasting

EGC Equal Gain Combining

ESNR Equivalent Signal-to-Noise plus Self-Interference Ratio

FDM Frequency Division Multiplexing

FDMA Frequency Division Multiple Access

FFT Fast Fourier Transform

FOBP Fractional Out-of-Band Power

FT Fourier Transform

HSOPA High Speed OFDM Packet Access

IBI Interblock Interference

ICI Interchannel Interference

IDFT Inverse Discrete Fourier Transform

IMP Intermodulation Product

ISI Intersymbol Interference

LINC Linear Amplification with Nonlinear Components

LTE Long-Term Evolution

MC-CDMA Multicarrier Code Division Multiple Access

MCM Multicarrier Modulation

MMSEC Minimum Mean Square Error Combining

MRC Maximum Ratio Combining

OFDM Orthogonal Frequency Division Multiplexing

OFDMA Orthogonal Frequency Division Multiple Access

ORC Orthogonality Restoring Combining

PDF Probability Density Function

PMEPR Peak-to-Mean Envelope Power Ratio

PSD Power Spectral Density

PSK Phase-Shift Keying

PTS Partial Transmit Sequence

QAM Quadrature Amplitude Modulation

QPSK Quadrature Phase-Shift Keying

RG Regular Grid

SC Single Carrier

SER Symbol Error Rate

SIR Signal-to-Interference Ratio

SNR Signal-to-Noise Ratio

SSPA Solid State Power Amplifier

VDSL Very High Bit-Rate Digital Subscriber Line

WLAN Wireless Local Area Network

WRAN Wireless Regional Area Network

ZP Zero-Padding

List of Symbols

General Symbols

$A(R)$ AM-to-AM conversion function

b number of transmitted bits

b_k number of transmitted bits on the kth subcarrier

C channel capacity

D_k kth frequency-domain nonlinearly distorted sample

D_k^{Qi} kth frequency-domain quantized sample

d minimum Euclidean distance between constellation points

$d(t)$ self-interference component of signal at nonlinearity output

$d_{\min,k}$ kth subcarrier minimum Euclidean distance between constellation points

d_n nth time-domain nonlinearly distorted sample

$d_n^{(2p+1)}$ nth time-domain distorted sample by nonlinear characteristic $f(R) = R^{2p+1}$

d_n^{Qi} nth time-domain quantized sample

ESNR_k kth subcarrier equivalent signal-to-noise plus self-interference ratio

$\overline{\text{ESNR}}$ equivalent signal-to-noise ratio geometric mean

F subcarrier separation

F_k kth frequency frequency-domain filtering coefficient

f frequency variable

$f(R)$ polar nonlinear characteristic function

f_0 carrier frequency/spectrum central frequency

f_c bandpass signal complex envelope reference frequency

f_p software radio pth channel central frequency

$G_d(f)$ power spectral density of nonlinearity self-interference

	component
$G_d^C(f)$	power spectral density of Cartesian nonlinearity self-interference component
$G_d^P(f)$	power spectral density of polar nonlinearity self-interference component
$G_s(f)$	power spectral density of transmitted signal
$G_s^{(p)}(f)$	power spectral density of the pth user transmitted signal
$G_{D,k}$	power spectrum of samples d_n (DFT of the autocorrelation $R_{d,n}$)
$G_{S,k}$	power spectrum of samples s_n (DFT of the autocorrelation $R_{s,n}$)
$G_{S,k}^C$	power spectrum of samples s_n^C (DFT of the autocorrelation $R_{s,n}^C$)
$G_{S,k}^C$	power spectrum of quantized samples s_n^Q (DFT of the autocorrelation $R_{s,n}^C$)
$G_{S,k}^{(p)}$	power spectrum of samples $s_n^{(p)}$ (DFT of the autocorrelation $R_{s,n}^{(p)}$)
$G_{Y,k}$	power spectrum of quantized received samples y_n^Q (DFT of the autocorrelation $R_{y,n}$)
$G_x(\tau)$	power spectral density of nonlinearity input signal
$G_x^C(\tau)$	power spectral density of a Cartesian nonlinearity input signal
$G_x^P(\tau)$	power spectral density of a polar nonlinearity input signal
$G_y(\tau)$	power spectral density of nonlinearity output signal
$G_y^C(\tau)$	power spectral density of a Cartesian nonlinearity output signal
$G_y^P(\tau)$	power spectral density of a polar nonlinearity output signal
$g(x)$	nonlinear characteristic function
$g_{\text{clip}}(x)$	clipping characteristic function
$g_{Qi}(x)$	quantization characteristic function
$g_{\text{quant}}(x)$	quantization characteristic function
H	channel gain
H_k	kth subcarrier overall channel frequency response
$H_k^{(p)}$	kth subcarrier overall channel frequency response for the pth user
$H_m(x)$	Hermite polynomial of degree m
$H^{(p)}(f)$	pth user channel impulse response
$H_T(t)$	frequency response of the reconstruction filter
$h(t)$	overall channel impulse response
$h_c(t)$	channel impulse response
$h_T(t)$	impulsive response of the reconstruction filter
I	average power of the self-interference component at the nonlinearity output

$I^{(2p+1)}$	average power of the self-interference component associated to nonlinear characteristic $f(R) = R^{2p+1}$
$\mathcal{I}$	set containing turned on subcarriers indexes
K	spreading factor
k	frequency index
$L_\gamma^{(1)}(x)$	generalized Laguerre polynomial of order γ
M	number of data symbols per block for each spreading code/user
$M_k^{(2p+1)}$	subcarrier multiplicity associated to nonlinear characteristic $f(R) = R^{2p+1}$
$M^{(2\gamma+1)}$	block of subcarrier multiplicities associated to nonlinear characteristic $f(R) = R^{2p+1}$
M_{Tx}	oversampling factor
m	data symbol index/constellation index
N	number of useful symbols/subcarriers
N'	number of symbols/subcarriers with oversampling factor
N_e	number of exponent bits of the quantizer
N_g	number of guard samples
N_k	channel noise for the kth frequency
N_m	number of mantissa bits of the quantizer
N_{on}	number of used subcarriers
n	time index
P	number of users/channels/spreading codes
$\overline{P}$	average power of transmitted signal
$P_{2\gamma+1}$	total power associated to the IMP of order $2\gamma+1$
$P_{2\gamma+1}^{(2p+1)}$	total power associated to the IMP of order $2\gamma+1$ for nonlinear characteristic $f(R) = R^{2p+1}$
$P_{2\gamma+1}^{Qi}$	total power associated to the IMP of order $2\gamma+1$ for a quantizer
p	user index
$p(x)$	probability density function
$p(x_1, x_2)$	joint probability density function
P_{out}	average power at the nonlinearity output
P_{out}^{Qi}	average power at the quantizer output
P_s	symbol error rate
$P_{s,k}$	kth subcarrier symbol error rate
$Q(x)$	error function
$R(t)$	complex envelope of complex signal
R_b	data rate
$R_d(\tau)$	autocorrelation of the self-interference component
$R_{d,n}$	autocorrelation of d_n
$R_{d,n}^{Qi}$	autocorrelation of d_n^{Qi}

$R_{s,n}$	autocorrelation of s_n
$R_{s,n}^C$	autocorrelation of s_n^C
$R_{s,n}^Q$	autocorrelation of s_n^Q
$R_{y,n}^Q$	autocorrelation of y_n^Q
$R_x(\tau)$	autocorrelation of nonlinearity input signal
$R_x^C(\tau)$	autocorrelation of Cartesian nonlinearity input signal
$R_x^P(\tau)$	autocorrelation of a polar nonlinearity input signal
$R_y(\tau)$	autocorrelation of nonlinearity output signal
$R_y^C(\tau)$	autocorrelation of a Cartesian nonlinearity output signal
$R_y^P(\tau)$	autocorrelation of a polar nonlinearity output signal
S	average power of the useful component at the nonlinearity output
$S^{(2p+1)}$	average power of the useful component associated to nonlinear characteristic $f(R) = R^{2p+1}$
$S(f)$	frequency-domain signal
$\tilde{S}(f)$	Fourier transform of $\tilde{s}(t)$
SIR_k	kth subcarrier signal-to-interference ratio
SIR_{Tx}	signal-to-interference ratio of the transmitted signal
S_k	kth frequency-domain transmitted symbol
S_k^C	kth frequency-domain nonlinearly distorted sample
$S_k^{(p)}$	kth frequency-domain/chip of the pth user/spreading code
S_k^Q	kth frequency-domain quantized sample
S_k^{Tx}	kth frequency-domain transmitted symbol with filtering procedure
$\tilde{S}_k$	kth frequency-domain data symbol
$\tilde{S}_{k,m}$	kth frequency-domain data symbol from mth burst
SNR	channel output signal-to-noise ratio
SNR^C	signal-to-noise ratio of the received signal
SNR_k	kth subcarrier signal-to-noise ratio
$\overline{\text{SNR}}$	signal-to-noise ratio geometric mean
$s(t)$	multicarrier burst/time-domain signal
$\tilde{s}(t)$	complex envelope of $s(t)$
$s_m(t)$	multicarrier mth burst
$\tilde{s}_m(t)$	complex envelope of $s_m(t)$
s_n	nth time-domain transmitted symbol
$\tilde{s}_n$	nth time-domain data symbol
$s_n^{(2p+1)}$	nth time-domain distorted samples by nonlinear characteristic $f(R) = R^{2p+1}$
s_n^C	nth time-domain nonlinearly distorted sample
s_n^Q	nth time-domain quantized sample
s_M	clipping level
$s_{MC}(t)$	complex envelope of multicarrier signal

T	useful part of burst duration
T_B	burst duration
T_G	guard period
t	time variable
$x(t)$	signal at input of nonlinearity
$x_{BP}(t)$	bandpass signal
$x_I(t)$	real part of signal $x(t)$
$x_Q(t)$	imaginary part of signal $x(t)$
Y_k	kth subcarrier received sample
$y(t)$	received signal/signal at output of nonlinearity
y_n	nth time-domain received sample
y_n^Q	nth time-domain received quantized sample
$y_p(t)$	software radio pth channel received signal
$y_{PF}(t)$	signal at the output of a polar nonlinearity
$W(f)$	Fourier transform of $w(t)$
$w(t)$	transmitted impulse/rectangular pulse
z_n	software radio nth time-domain transmitted symbol
z_n^Q	nth time-domain quantized sample
$z(t)$	software radio transmitted signal
$\tilde{z}(t)$	complex envelope of software radio transmitted signal

Greek Letters Symbols

α	scale factor associated to the nonlinear operation
$\alpha^{(2p+1)}$	scale factor associated to nonlinear operation $f(R) = R^{2p+1}$
α^C	scale factor associated to a Cartesian nonlinear operation
α_k	parameter depending on adopted constellation
α^P	scale factor associated to a polar nonlinear operation
α^{Qi}	scale factor associated to a quantizer operation
β_{cross}	cross QAM constellation scale factor
β_k	parameter depending on adopted constellation
β_m	coefficients of power series expansion
β_{square}	square QAM constellation scale factor
Γ	SNR gap
γ	index associated to the IMP
γ_c	coding gain
γ_m	system margin
Δf	frequency shift
ΔN	number of shifted samples
ε	symbol energy

ε_k kth subcarrier symbol energy

$\varepsilon_{\mathrm{tot}}$ total energy

η_G degradation factor associated to the use of guard bands

$\Theta(R)$ AM-to-PM conversion function

ξ_p pth user power control weighting coefficient

ρ normalized autocorrelation

σ^2 variance of transmitted frequency-domain symbols

σ_N^2 channel noise variance

$\sigma_{N,k}^2$ kth subcarrier channel noise variance

σ_p^2 pth user transmitted frequency-domain data symbols variance

σ_S^2 transmitted frequency-domain data symbols variance

$\phi(t)$ argument of the complex signal

Ψ_p set of frequencies associated to the pth user

$\Psi(f_1, f_2)$ joint characteristic function

Common Operators

$\delta(f)$ Dirac delta function

$E[\cdot]$ ensemble average

$\mathcal{F}\{\cdot\}$ Fourier transform

$\mathrm{Im}\{\cdot\}$ imaginary part

$\mathrm{Re}\{\cdot\}$ real part

Chapter 1

Introduction

With the growing demand for high-speed data transmission services, future communication systems are required to support reliable, high quality of service, as well as spectral and power efficiency. New services such as video streaming, interactive multimedia communications, video on demand, e-commerce, digital telephony, and digital radio require larger and larger bandwidth for end users.

Multicarrier Modulation (MCM) systems have found increasing use in these modern communication services, including both wireline and wireless environments, due to their ability to fulfill most of the requirements of such systems. Some of the most well-known applications for multicarrier modulations are Digital Subscriber Line (DSL) [Bin00], over wired media using Discrete Multitone (DMT), and Digital Audio Broadcasting (DAB) [ETS06], Digital Video Broadcasting (DVB) [ETS09] and Wireless Local Area Networks (WLANs) [vNP00] for wireless communications using Orthogonal Frequency Division Multiplexing (OFDM). The variant Orthogonal Frequency Division Multiple Access (OFDMA) is used in wireless broadband access technologies IEEE 802.16a/d/e (commonly referred to as WiMAX [IEE04, IEE06]) and 3rd Generation Partnership Project (3GPP) Long-Term Evolution (LTE) (named High Speed OFDM Packet Access (HSOPA)) [3GP06]. It is also the access method candidate for Wireless Regional Area Networks (WRANs) [IEE].

The basic principle of multicarrier modulation is to transmit data by splitting it into several components and sending each component over separate carrier signals. This means that the available channel bandwidth is divided into a number of equal-bandwidth subchannels, where

1

the bandwidth of each subchannel is narrow enough for the frequency response characteristics of the subchannels to be nearly ideal.

Advantages of multicarrier modulations include low computational complexity implementation by using a Discrete Fourier Transform (DFT)/Inverse Discrete Fourier Transform (IDFT) pair (which can be easily performed by means of the Fast Fourier Transform (FFT) algorithm), relative immunity to fading caused by transmission over more than one path at a time (multipath fading), less susceptibility than single-carrier systems to interference caused by impulse noise and enhanced immunity to inter-symbol interference. Limitations include difficulty in time synchronization, power, and spectral efficiency loss due to the guard interval and sensitivity to frequency offset. Another major drawback is their vulnerability to nonlinear distortion effects, which come from the fact that most of the components of the transmitter and receiver do not show perfectly linear behaviours. These include the DFT/IDFT pair, the high power amplifier and the digital-to-analog and analog-to-digital converters, among others. Consequently, developments in the area of modeling and analysis of multicarrier communication systems in the presence of nonlinearities are required.

The development of analytical techniques to assess the impact of nonlinearities on the performance of multicarrier signals has been the subject of extensive research. However, the fact that such techniques focus on the performance of signals distorted by specific nonlinear devices, such as clipping devices or high-power amplifiers, causes dispersion of techniques and results.

To predict the degradation introduced by a nonlinearity, as well as to evaluate the performance of multicarrier systems where they occur, it is usual to resort to Monte Carlo simulations, which require large computation times. Another approach takes advantage of the Gaussian-like nature of these signals when the number of subcarriers is high (say, greater than 64). This nature can be used to analytically characterize a multicarrier signal submitted to a nonlinear device and this characterization can then be employed for performance evaluation of nonlinearly distorted multicarrier signals [BC00, DTV00, DG04].

Since practical multicarrier schemes have a finite number of subcarriers, the Gaussian approximation for multicarrier signals might not be accurate, especially when the number of subcarriers is not very high. In fact, by taking advantage of the asymptotic properties of the large excursions of a stationary Gaussian process, it can be shown that the errors of the Gaussian approximation might be significant [BSGS02]. A study of the correctness of this approximation to quantify the distortion caused by nonlinear devices is, therefore, critical to the future of multicarrier techniques.

The main motivation behind the development of analytical techniques for the evaluation of nonlinear distortion effects is its application to problems that arise in different multicarrier schemes.

As mentioned earlier, one of the sources of nonlinear distortion on multicarrier systems is the DFT/IDFT pair used in the transmitter and the receiver. The numerical accuracy inherent to these operations nonlinearly distorts the signal, in an effect known as quantization, which can have a significant impact on the transmission performance. This accuracy can be modelled by including appropriate quantization devices in the transmission chain, thus allowing a statistical characterization and evaluation of given quantization characteristics.

The software radio concept is a topic of widespread interest in wireless cellular systems. A software radio base station uses an Analog-to-Digital Converter (ADC) that has to be able to sample a wideband signal associated to the combination of a large number of users, possibly with different bands and different powers. The high sampling ratio and quantization requirements make the ADC a key component of any software radio architecture. The development of an analytical tool that enables the evaluation and optimization of the quantization requirements within the ADC is therefore a topic of great interest.

One of the advantages of multicarrier modulation is its ability to adapt the subcarrier operating parameters to the communications channel, thereby enhancing the overall performance of the system. Although several subcarrier parameters are available for adjustment, one of the most popular choices is subcarrier signal constellation size. The process of adjusting this parameter, called adaptive bit loading, involves an algorithm that adjusts the number of bits per subcarrier (and corresponding constellation size) according to the channel conditions, i.e., the transmitted number of bits is not equal across all subcarriers [HH87, LC97, KRJ00, LSC07]. Another parameter that can be adjusted is the energy assigned to each subcarrier. It can be chosen according to the number of bits and channel attenuation on a particular subcarrier. Consequently, adaptive bit loading has the potential to achieve data transmissions that are very spectrally efficient. These algorithms can be combined with analytical techniques to evaluate nonlinear distortion effects and used to study the impact of these effects on bit and energy distributions, thus obtaining more reliable systems.

OFDMA is a type of multicarrier scheme that is used in several wireless broadband access technologies. As other multicarrier modulations, the transmitted signals have large envelope fluctuations, leading to amplification difficulties and making them very prone to nonlinear distortion effects [GV94, CWETM98]. For this reason, several techniques have been proposed to reduce the envelope fluctuations of these signals, namely through suitable preprocessing schemes. These techniques (for example, involving clipping) can also significantly nonlinearly distort the signal. As a consequence, the performance evaluation of OFDMA systems in the presence of nonlinear effects (either due to an imperfectly linear amplification or to suitable signal processing schemes to reduce the envelope fluctuations of the transmitted signals) is a relevant issue for the implementation of future OFDMA systems.

This book is dedicated to the analysis of nonlinear distortion of multicarrier signals. It aims to present an unified approach to results on nonlinear distortion effects on Gaussian signals, to study the accuracy of the usual Gaussian approximation used for its evaluation and to develop analytical techniques to investigate the impact of some nonlinear devices on several multicarrier systems.

1.1 Book Structure

After this brief introduction, Chapter 2 is dedicated to a short overview of the transmission techniques considered in this thesis, presenting the basic principles of multicarrier modulations. OFDM modulations and other multicarrier schemes are also described, including the characterization of the transmitted signals in the time and frequency domains and the transmitter and receiver structures.

Subsequently, in Chapter 3, a review of well-known results on nonlinear distortion of Gaussian signals is presented. The presented mathematical analysis is based on the fact that when the number of subcarriers is high enough, multicarrier signals can be assumed to be complex-Gaussian distributed. Memoryless nonlinear devices, Cartesian memoryless devices and polar memoryless devices, all with Gaussian inputs, are considered.

Chapter 4 is dedicated to the study of the accuracy of the Gaussian approximation generally used for the evaluation of nonlinear effects. An exact characterization of nonlinearly distorted multicarrier signals submitted to nonlinear devices with characteristics of the form R^3 and R^5 is presented and generalized for higher order powers. The results obtained using this characterization are then compared with the ones obtained using the Gaussian approximation.

Chapter 5 is devoted to the application of the nonlinear distortion analysis techniques presented in Chapter 3 to some of the multicarrier systems described in Chapter 2. In all cases the developed methods take advantage of the Gaussian behaviour of the complex envelope of multicarrier signals. Section 5.1 is dedicated to clipping and filtering techniques to reduce the well-known high envelope fluctuations of multicarrier signals. In Section 5.2, quantization effects caused by the numerical accuracy required in the DFT/IDFT operations are analyzed using an appropriate statistical characterization of the signal along the transmission chain. Section 5.3 presents an analytical tool for evaluating the quantization requirements within the ADC used in software radio architectures. In Section 5.4, loading techniques are redefined to take into account nonlinear distortion issues and used to study and improve the system margin and subcarrier bit and energy distribution of adaptive multicarrier schemes. A study of the impact of nonlinear distortion effects inherent to nonlinear signal processing techniques for reducing the Peak-to-Mean Envelope Power Ratio (PMEPR) of the transmitted signals in OFDMA systems is presented in Section 5.5. Evaluation of both uplink and downlink transmission is included. For all these applications, a set of performance results is presented and discussed.

Chapter 2

Multicarrier Transmission

The basic principle of multicarrier transmission is to split the available bandwidth into several smaller frequency bands. The main data stream is separated into many low-rate substreams that are transmitted in parallel on different frequency channels, i.e., data transmission is performed independently in different subcarriers.

The principle of dividing data in multiple sequences and to use them to modulate a subcarrier was originally applied in [DHM57]. Since then, various designations have been used, depending on deployment and application: orthogonally multiplexed Quadrature Amplitude Modulation (QAM) [HHS86], dynamically assigned multiple QAM [KB80], Discrete Multitone (DMT) [Cio91], synchronized DMT [JBC95], Orthogonal Frequency Division Multiplexing (OFDM) [Cim85], etc.

Motivations for using this technique include no need for a complex equalization and its high spectral efficiency. This spectral efficiency is achieved despite the use of guard bands by using a sufficiently large number of subcarriers with a relatively narrow bandwidth each, and by allocating the transmitted power, constellation size and coding rate to each subcarrier efficiently. Another reason for the interest in this type of modulation is its ability to cope with severely time dispersive channels without the need of complex receiver implementations [Bin90, Sal67, WE71]. A simple way to implement multicarrier modulation, suggested in [WE71], uses a pair of Discrete Fourier Transforms (DFTs) in the transmitter and the receiver, which can be easily performed using the Fast Fourier Transform (FFT) algorithm [CT65].

In recent years, multicarrier modulation schemes have been successfully used in several digital transmission systems, such as OFDM schemes in Wireless Local Area Networks (WLANs) [vNP00] and DMT in Asymmetric Digital Subscriber Line (ADSL) [Bin00]. They were also selected for several broadband wireless systems, such as Digital Video Broadcasting (DVB) [ETS09], wireless broadband access technologies IEEE 802.16a/d/e (commonly referred to as WiMAX [IEE04, IEE06]) and 3GPP (3rd Generation Partnership Project) Long-Term Evolution (LTE), named High Speed OFDM Packet Access (HSOPA) [3GP06]. They are also the access method candidate for Wireless Regional Area Networks (WRANs) [IEE].

In this chapter, a short overview of different multicarrier schemes is presented, including their basic principles, characterization of the transmitted signals and transmitter and receiver structures.

2.1 Orthogonal Multicarrier Signals

The complex envelope of a multicarrier signal can be written as a sum of bursts transmitted at rate $R_b = 1/T_B$, i.e.,

$$s_{\mathrm{MC}}(t) = \sum_m s_m(t - mT_B), \qquad (2.1)$$

where T_B is the burst duration and $s_m(t)$ denotes the mth burst, which can be written as

$$s_m(t) = \sum_{k=0}^{N-1} |\tilde{S}_{k,m}| \, \cos(2\pi(f_0 + kF)t + \arg(\tilde{S}_{k,m})) \, w(t), \qquad (2.2)$$

where N is the number of subcarriers, $w(t)$ represents the transmitted impulse, f_0 is the carrier frequency, F is the subcarriers separation and $\tilde{S}_{k,m} = |\tilde{S}_{k,m}| \, e^{j \arg(\tilde{S}_{k,m})}$ represents the complex symbol transmitted on the kth subcarrier of the mth burst. This symbol is selected from a given constellation (e.g., Phase-Shift Keying (PSK) or QAM constellations), according to the transmitted data and the subcarriers channel response, i.e., different constellations can be assigned to different subcarriers. Clearly, $s_m(t)$ can be written as

$$s_m(t) = \mathrm{Re} \{\tilde{s}_m(t)\}, \qquad (2.3)$$

with

$$\tilde{s}_m(t) = \sum_{k=0}^{N-1} \tilde{S}_{k,m}\, e^{j2\pi(f_0+kF)t}\, w(t), \qquad (2.4)$$

i.e., $\tilde{s}_m(t)$ is the complex envelope of $s_m(t)$ referred to frequency f_0 (we assume that f_0 corresponds to the central frequency of the spectrum, unless otherwise stated). By applying the Fourier Transform (FT), defined as

$$S(f) \triangleq \mathcal{F}\{s(t)\} = \int_{-\infty}^{+\infty} s(t)\, e^{-j2\pi ft} dt, \qquad (2.5)$$

to both sides of (2.4), we obtain

$$\tilde{S}(f) = \sum_{k=0}^{N-1} \tilde{S}_k\, W(f - kF), \qquad (2.6)$$

with $W(f) = \mathcal{F}\{w(t)\}$ (for the sake of simplicity, we omit the burst index m). For a given burst, information symbols are sequentially transmitted in the frequency domain, each one on a different subcarrier, being all transmitted on a time interval of duration T. Note that, if the bandwidth associated to $W(f)$ is smaller then $F/2$, then the transmission band of each symbol is a fraction of order $1/N$ of the total transmission band. The simplest case of multicarrier modulation is conventional Frequency Division Multiplexing (FDM) scheme, where the spectrum associated with different symbols does not overlap. If we assume that the bandwidth associated to $W(f)$ is smaller then $F/2$, then each symbol $\tilde{S}_k$ occupies a fraction $1/N$ of the total transmission band, as shown in Figure 2.1.

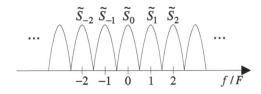

Figure 2.1: Conventional FDM.

In conventional Single Carrier (SC) schemes, the complex envelope of a N-symbol burst can be written as

$$s_{\text{SC}}(t) = \sum_{k=0}^{N-1} \tilde{S}_k\, w(t - kT), \qquad (2.7)$$

where $w(t)$ represents the transmitted pulse, T is the symbol duration and $\tilde{S}_k$ denotes the kth symbol.

Note that (2.7) is a dual version of (2.6), i.e., there is a duality relation between conventional single carrier and multicarrier modulations, with time and frequency domains exchanging roles. This duality is also present in the orthogonality conditions between pulses and subcarriers, respectively, for each of the modulation schemes. For conventional single carrier modulations the orthogonality condition between pulses is

$$\int_{-\infty}^{+\infty} w(t - kT)\, w^*(t - k'T)dt = 0, \qquad k \neq k', \tag{2.8}$$

(* denotes 'complex conjugate'), which ensures an Intersymbol Interference (ISI) free transmission. For multicarrier modulation, the orthogonality condition between subcarriers is

$$\int_{-\infty}^{+\infty} W(f - kF)\, W^*(f - k'F)df = 0, \qquad k \neq k'. \tag{2.9}$$

Using Parseval's theorem, this condition can be rewritten as

$$\int_{-\infty}^{+\infty} |w(t)|^2\, e^{-j2\pi(k-k')Ft}dt = 0, \qquad k \neq k', \tag{2.10}$$

which guarantees there is no Interchannel Interference (ICI). Note that (2.8) is the dual version of (2.9).

It is well-known that with linear single carrier modulations, the impulses $w(t - kT)$, $k \in \mathbb{Z}$, can be orthogonal even if they overlap in time. Similarly, in multicarrier modulation it is possible to verify the orthogonal condition (2.9) even if functions $W(f - kF)$, $k = 0, \ldots, N - 1$, overlap in the frequency domain. For example, for single carrier modulations, impulse

$$w(t) = \text{sinc}\left(\frac{t}{T}\right), \tag{2.11}$$

with $\text{sinc}(x) \triangleq \sin(\pi x)/(\pi x)$, satisfies condition (2.8). Hence, for multicarrier modulations, condition (2.9) can be satisfied by

$$W(f) = \text{sinc}\left(\frac{f}{F}\right). \tag{2.12}$$

In this case, $w(t)$ is a rectangular pulse of duration $T = 1/F$, for example,

$$w(t) = \begin{cases} 1, & t_0 \leq t \leq t_0 + T \\ 0, & \text{otherwise}, \end{cases} \tag{2.13}$$

and (2.9) reduces to

$$\int_{-\infty}^{+\infty} |w(t)|^2 e^{-j2\pi(k-k')Ft}dt = \int_{t_0}^{t_0+T} e^{-j2\pi(k-k')Ft}dt$$

$$= \begin{cases} T, & k = k' \\ 0, & \text{otherwise,} \end{cases} \qquad (2.14)$$

despite $W(f - kF) = T\text{sinc}((f - kF)T)$ and $W(f - k'F) = T\text{sinc}((f - k'F)T)$ not being disjoint in the frequency domain.

As will be seen in Subsection 2.2.2, in time dispersive channels bursts overlap, due to multipath propagation. To deal with this problem, it is usual to adopt a guard interval between symbols. If the length of this interval is longer than the overall impulse response (which includes the channel impulse response and the transmitter and detection filters) then there is no ISI nor ICI. This can be achieved by extending each burst with any known fixed sequence, for example, a cyclix prefix (CP), an all-zero sequence zero-padding (ZP), or a pseudonoise (PN) sequence, usually denoted as PN extension or unique word (UW). The most popular is CP, but ZP schemes can be a good alternative to CP-assisted schemes [MCDG00]. However, complex receiver structures must be employed, involving the inversion or multiplication of matrixes whose dimensions grow with the block length. Employing overlap-and-add techniques [MCG+00] allows the receiver complexity to be similar to the conventional CP-assisted schemes, but performance is also identical. Efficient FFT-based receivers can also be designed for ZP-OFDM [AD06, DSA09]. To avoid power and spectral degradation, the guard interval must be a small fraction of the overall length of the blocks. In this work, we will only consider CP-assisted block transmission.

2.2   Orthogonal Frequency Division Multiplexing

In Orthogonal Frequency Division Multiplexing (OFDM), one of the most used particular case of multicarrier modulation, adopted pulses are of the form

$$w(t) = \begin{cases} 1, & -T_G \le t \le T \\ 0, & \text{otherwise,} \end{cases} \qquad (2.15)$$

i.e., it is a rectangular pulse with duration $T_B = T + T_G$. Although this impulse doesn't verify condition (2.9), we can say that the different subcarriers are orthogonal in the interval $[0, T]$. In fact,

$$\int_0^T |w(t)|^2 \, e^{-j2\pi(k-k')Ft} dt = \int_0^T e^{-j2\pi(k-k')Ft} dt = 0, \qquad k \neq k'.$$

$$(2.16)$$

Interval $[0, T]$ will be used by the receiver as the effective detection interval.

2.2.1 Transmitter Structure

Given equation (2.4), the multicarrier burst can be written as

$$\tilde{s}(t) = s^{(P)}(t) \, w(t), \tag{2.17}$$

where

$$s^{(P)}(t) = \sum_{k=0}^{N-1} \tilde{S}_k \, e^{j2\pi kFt}. \tag{2.18}$$

Note that, since $s^{(P)}(t)$ is a time-domain periodic function with period T, the complex envelope associated with the guard period is a repetition of the last part of the transmitted burst, as illustrated in Figure 2.2, i.e.,

$$\tilde{s}(t) = \tilde{s}(t+T), \qquad -T_G \leq t \leq 0. \tag{2.19}$$

This guard period can be viewed as an artificial extension of the burst and is usually designated by CP. This extension is removed by the receiver, as will be seen in Subsection 2.2.2.

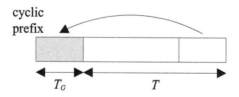

Figure 2.2: Cyclic prefix.

By applying the FT to both sides of (2.18), we obtain

$$S^{(P)}(f) = \mathcal{F}\{s^{(P)}(t)\} = \sum_{k=0}^{N-1} \tilde{S}_k \, \delta\left(f - \frac{k}{T}\right), \tag{2.20}$$

where $\delta(f)$ represents the Dirac delta function. From this equation, it is clear that $s^{(P)}(t)$ occupies a band of width N/T, hence, according to the sampling theorem, it can be completely recovered from its samples taken in the interval $[0, T]$ at sample rate N/T. These samples can be written as

$$s_n^{(P)} \triangleq s^{(P)}(t)\Big|_{t=\frac{nT}{N}} = \sum_{k=0}^{N-1} \tilde{S}_k \, e^{j2\pi\frac{kn}{N}}, \qquad n = 0, 1, \ldots, N-1. \quad (2.21)$$

Noting that the Inverse Discrete Fourier Transform (IDFT) of the block $\{X_k; k = 0, 1, \ldots, N-1\}$ is the block $\{x_n; n = 0, 1, \ldots, N-1\}$, with

$$x_n = \frac{1}{N} \sum_{k=0}^{N-1} X_k \, e^{j2\pi\frac{kn}{N}}, \qquad (2.22)$$

and defining

$$\tilde{s}_n \triangleq \frac{1}{N} \, s_n^{(P)}, \qquad (2.23)$$

it is clear that the block of time domain samples $\{\tilde{s}_n; n = 0, 1, \ldots, N-1\}$ is the IDFT of the block of data symbols $\{\tilde{S}_k; k = 0, 1, \ldots, N-1\}$, with

$$\tilde{S}_k = \begin{cases} \tilde{S}_k, & 0 \leq k \leq N/2 - 1 \\ \tilde{S}_{k-N}, & N/2 \leq k \leq N-1. \end{cases} \qquad (2.24)$$

This means that the sampled version of $s^{(P)}(t)$ in the interval $[0, T]$ can be obtained simply by calculating the IDFT of the block of transmitted symbols, which can be efficiently implemented using the well-known FFT algorithm [CT65] when N is a power of 2.

After generating the samples of $s^{(P)}(t)$, the wave shape associated to a given burst is obtained multiplying those samples by the samples of the time window $w(t)$ (whose time duration is longer than T), i.e., the transmitted samples are $\tilde{s}_n \tilde{w}_n$, with

$$\tilde{w}_n \triangleq w\left(\frac{nT}{N}\right). \qquad (2.25)$$

Finally, the analog signal associated to a given burst is generated from these samples by digital to analog conversion and reconstruction filtering (see Figure 2.6). Its complex envelope can be written as

$$\tilde{s}(t) = \left(\left(\sum_{n=-\infty}^{+\infty} \tilde{s}_n \, \delta \left(t - n\frac{T}{N} \right) \right) \cdot w(t) \right) * h_T(t)$$

$$= \left(\sum_{n=-\infty}^{+\infty} \tilde{s}_n \, \tilde{w}_n \, \delta \left(t - n\frac{T}{N} \right) \right) * h_T(t)$$

$$= \sum_{n=-\infty}^{+\infty} \tilde{s}_n \, \tilde{w}_n \, h_T \left(t - n\frac{T}{N} \right), \qquad (2.26)$$

where $h_T(t)$ is the impulsive response of the reconstruction filter.

To simplify the reconstruction filter used in the reception, it is usual to sample the burst at sampling rate $M_{\text{Tx}} N/T$ instead of N/T, i.e., to use an oversampling factor M_{Tx} greater than one and not necessarily integer. Usually the original block $\{\tilde{S}_k; k = 0, 1, \ldots, N-1\}$ already includes $2N_z$ null symbols (that correspond to 'idle' subcarriers), half at the beginning and the other half at the end of the burst. This is equivalent to oversampling the burst by a factor $N/(N - 2N_z)$, with $N - 2N_z$ useful subcarriers.

For an oversampling factor of M_{Tx} and a reference burst with N subcarriers, the samples of $s^{(P)}(t)$ in the interval $[0, T]$ are given by

$$s_n^{M_{\text{Tx}}} \triangleq s^{(P)}(t) \Big|_{t=\frac{nT}{N'}} = \sum_{k=0}^{N-1} \tilde{S}_k \, e^{j2\pi \frac{kn}{N'}}, \qquad n = 0, 1, \ldots, N'-1, \quad (2.27)$$

where $N' = M_{\text{Tx}} N$. As before, defining

$$s_n \triangleq \frac{1}{N'} s_n^{M_{\text{Tx}}}, \qquad (2.28)$$

it is clear that, apart from the scalar factor $1/N'$, the block of time domain samples $\{s_n; n = 0, 1, \ldots, N'-1\}$ is the IDFT of the zero-padded block $\{S_k; k = 0, 1, \ldots, N'-1\}$, with

$$S_k = \begin{cases} \tilde{S}_k, & 0 \leq k \leq N/2 - 1 \\ 0, & N/2 \leq k \leq N' - N/2 - 1 \\ \tilde{S}_{k-N}, & N' - N/2 \leq k \leq N' - 1. \end{cases} \qquad (2.29)$$

Once again, this means that the oversampled version of $s^{(P)}(t)$ can be obtained simply by calculating the IDFT of the extended block.[1]

---

[1]Throughout this work, we will designate by in-band subcarriers the N subcarriers that correspond to the reference burst, i.e., the ones that might carry useful information, and by out-of-band subcarriers the $N' - N$ idle subcarriers associated to the oversampling.

The complex envelope of the analog signal associated to a multicarrier burst with an oversampling factor M_{Tx} can be written as

$$s(t) = \sum_{n=-\infty}^{+\infty} s_n \, w_n \, h_T \left(t - n\frac{T}{N'} \right),$$

(2.30)

with

$$w_n \triangleq w \left(\frac{nT}{N'} \right).$$

(2.31)

This signal does not exactly match the representation of the multicarrier burst given by (2.17), but the difference is small, especially for a large number of subcarriers or a high oversampling factor, with differences occurring mainly on the extremes of the interval occupied by $w(t)$. When $M_{\text{Tx}} \to \infty$ the signal given by equation (2.30) approaches the reference burst (2.17).

To reduce the out-of-band radiation levels on the spectrum of multi-carrier bursts, it is common to employ a square-root raised-cosine window for $w(t)$, instead of a rectangular shape window [DG04]. In this case, the signal associated with a given burst still has complex envelope given by (2.30), with

$$w(t) = w'(t) * h_W(t),$$

(2.32)

where

$$h_W(t) = \frac{\pi}{2T_W} \cos \left(\frac{\pi t}{T_W} \right) \text{rect} \left(\frac{t}{T_W} \right)$$

(2.33)

and $w'(t)$ is a rectangular pulse with duration $T_B = T + T_G + T_W$. As shown in Figure 2.3, the duration of the time window $w(t)$ is $T_B + T_W = T + T_G + 2T_W$, which results in an overlap of T_W between adjacent bursts. This means that this raised-cosine window has roll-off factor $\frac{T_W}{T_B}$. In this work, we assume $T_W = 0$.

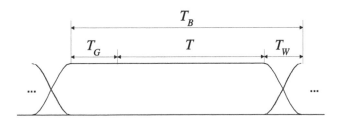

Figure 2.3: Raised-cosine time window.

In conventional multicarrier implementations, the complex envelope of the signal is referred to the central frequency of the spectrum $f_0 = \frac{N'}{2T}$, as depicted in Figure 2.4.A. Alternatively, it can be referred to a different frequency $f_c = f_0 + \Delta N F$, leading to a shift in the spectrum of the signal, as depicted in Figure 2.4.B. In this case (2.18) is substituted by

$$s^{(P)}(t) = \sum_{k=-\Delta N}^{N-\Delta N-1} S_k^{\Delta N} e^{j2\pi kFt},\tag{2.34}$$

where samples $S_k^{\Delta N}$ are simply obtained by shifting ΔN samples of the extended block given by (2.29), as illustrated in Figure 2.5. The signal

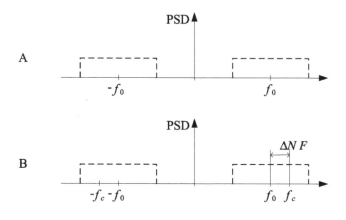

Figure 2.4: Complex envelope referred to the central frequency of the spectrum f_0 (A) and to f_c (B).

still occupies a bandwidth of $\frac{N'}{2T}$, hence, it can be generated from a sampled version of (2.34) with sampling frequency $\frac{N'}{T}$. The corresponding samples are

$$s_n^{M_{\text{Tx}}} \triangleq \tilde{s}(t)\big|_{t=\frac{nT}{N'}} = \sum_{k=-\Delta N}^{N-\Delta N-1} S_k^{\Delta N} e^{j2\pi \frac{kn}{N'}}, \qquad n = 0, 1, \ldots, N'-1,\tag{2.35}$$

and we can define the transmitted samples s_n as in (2.28).

The block diagram of multicarrier transmitter employing an IDFT is shown in Figure 2.6.

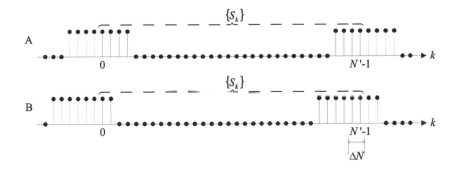

Figure 2.5: Block $\{S_k^{\Delta N}\}$ with $N = 8$, $N' = 4N$ and $\Delta N = 0$ **(A)** or $\Delta N = N/4$ **(B)**.

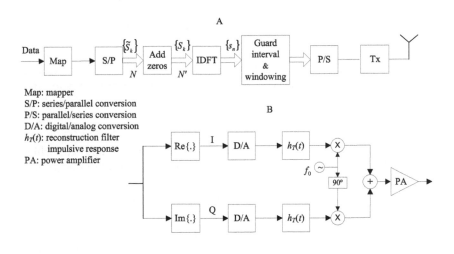

Map: mapper
S/P: series/parallel conversion
P/S: parallel/series conversion
D/A: digital/analog conversion
$h_T(t)$: reconstruction filter
       impulsive response
PA: power amplifier

Figure 2.6: OFDM transmitter structure **(A)** and detail of the 'Tx' block **(B)**.

2.2.2   Receiver Structure

Figure 2.7 shows the multicarrier receiver structure. Just as the signal generation on the transmitter can be performed using an IDFT, signal detection on the receiver can be implemented using a DFT. Although not necessary, an oversampling factor is assumed on the reception, equal to the one on the transmitter, that is, the receiver's sample rate is N'/T.

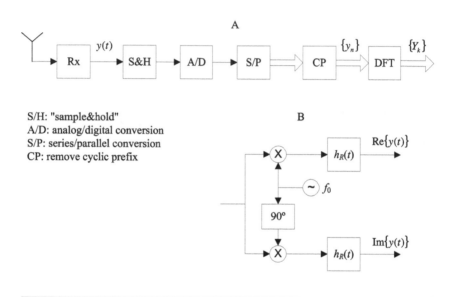

Figure 2.7: OFDM receiver structure (A) and detail of the 'Rx' block (B).

The samples corresponding to the cyclic prefix are ignored, which means that only the samples corresponding to an interval of duration T are used. We assume the detection filter $h_R(t)$ is adapted to the transmitter filter $h_T(t)$, that is, its frequency response is shaped like a square-root raised-cosine with bandwidth $\frac{N'}{2T}(1 + \rho)$.

Due to multipath propagation, the received bursts will overlap, as illustrated in Figure 2.8. Moreover, the different subcarriers of the same burst will interfere with each other. However, this problem can be bridged in the following way: the detection operates only on the samples associated to an useful interval of duration T, that is, if the samples corresponding to the cyclic prefix are ignored. If the duration of the cyclic prefix T_G is longer than the overall Channel Impulse Response (CIR) (which includes the impact of the transmission and detection filters, as well as the channel itself), the overlapping of the bursts is restricted to the guard interval, thus preventing the effects of the bursts overlap on the samples associated with the useful interval. This is usually referred as 'absence of ISI', although probably it would be more adequate to use the term interblock interference (IBI).

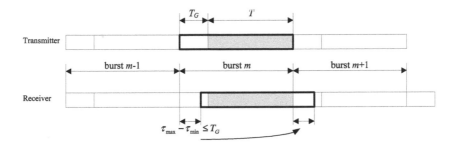

Figure 2.8: Intersymbol interference (ISI) elimination using guard intervals.

Let $h_c(t)$ denote the channel impulse response. The overall channel impulse response is characterized by $h(t) = h_T(t) * h_c(t) * h_R(t)$. If both the samples S_k and the corresponding time domain samples s_n are periodic, then the received sample at the kth subcarrier is given by [BS02, SCD$^+$10]

$$Y_k = H_k S_k + N_k, \qquad k = 0, 1, \ldots, N' - 1, \qquad (2.36)$$

where H_k is the overall channel frequency response for the kth subcarrier and is given by

$$H_k = \frac{N'}{T} H_T\left(\frac{k}{T}\right) H_c\left(\frac{k}{T}\right) H_R\left(\frac{k}{T}\right), \qquad k = 0, 1, \ldots, N' - 1, \quad (2.37)$$

and N_k is the channel noise component for that subcarrier, which we assume statistically independent and Gaussian distributed with zero mean and variance $N_0(N')^2/T$ on both the in-phase and quadrature components. Assuming that $H_T(f) = H_R(f) = 1$ for the N useful subcarriers, and that the reception filter totally eliminates the aliasing effects, we can write

$$H_k = \frac{N'}{T} H_c\left(\frac{k}{T}\right), \qquad k = 0, 1, \ldots, N' - 1. \qquad (2.38)$$

Note that the channel simply acts like a multiplicative factor on each subcarrier, thus preserving orthogonality between subcarriers in the useful interval. This is usually referred to as 'absence of ICI'. This is due to the fact that the signal is a cyclic prefix of each burst on the guard interval. Hence, in the useful interval T, as illustrated in Figure 2.9, everything is processed as if the transmitted signal is not the sequence of bursts expressed by equation (2.1), but the periodic signal $s^{(P)}(t)$ given by (2.18) corresponding to $s(t)$. Therefore, the linear convolution associated with the channel is formally equivalent to a circular convolution with respect to the useful part of the multicarrier block. The absence of

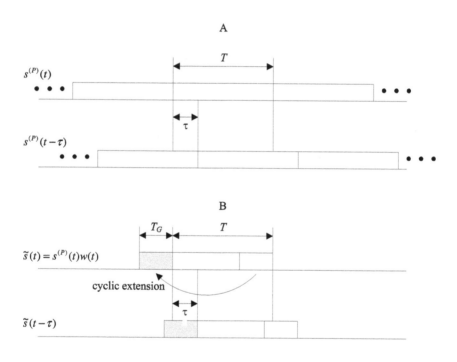

Figure 2.9: Impact of a dispersive channel on the signal $s^{(P)}(t)$ (A) and on the corresponding burst (B).

ISI and ICI means that a multicarrier system can be viewed as a set of N non-dispersive parallel channels, as illustrated in Figure 2.10.

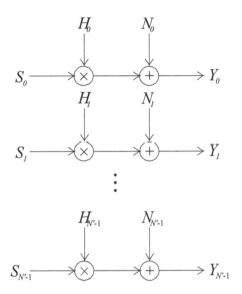

Figure 2.10: Independent parallel channels.

2.3   Other Multicarrier Schemes

2.3.1   Discrete Multitone

Discrete Multitone (DMT) modulation is the multicarrier technique with applications in high data rate transmission through twisted-pair copper wires, namely Asymmetric Digital Subscriber Line (ADSL) and Very High Bit-Rate Digital Subscriber Line (VDSL) systems. The DMT signal can be seen as a baseband OFDM modulation or as a bandpass signal with carrier frequency of magnitude order close to the bandwidth. These two approaches of looking at a DMT signal suggest different transmitter structures, which will be described next.

2.3.1.1   Bandpass Approach

When the signal is seen as a bandpass signal its complex envelope is referred to frequency f_c, i.e., we can write

$$s(t) = \mathrm{Re}\left\{\tilde{s}(t)\, e^{j2\pi f_c t}\right\}, \tag{2.39}$$

where the complex envelope of the bandpass signal $s(t)$ is given by (2.17), with $s^{(P)}(t)$ given by (2.34). The corresponding samples are given by (2.35). The structure of the bandpass DMT transmitter and receiver are shown, respectively, in Figures 2.11 and 2.12.

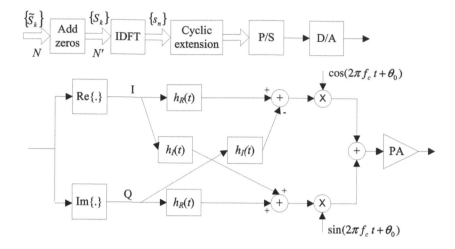

Figure 2.11: Bandpass DMT transmitter structure.

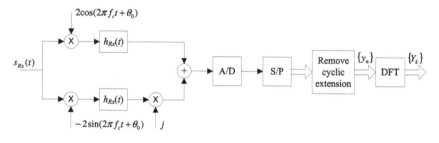

Figure 2.12: Bandpass DMT receiver structure.

Note that if f_c is different from the central frequency of the bandpass spectrum f_0, then the frequency response from the reconstruction filter is asymmetric and, consequently, its impulse response is complex, i.e., $h_T(t) = h_R(t) + jh_I(t)$, with $h_R(t)$ and $h_I(t)$ real. This means we will need 4 reconstruction filters, as shown in Figure 2.11. If $f_c = f_0$, then $h_I(t) = 0$ and only two filters will be necessary. We also note that the two approaches have similar implementation complexity. The baseband structure requires a FFT with $N' = 2M_{\text{Tx}}N$ points, with corresponding real samples. Bandpass structure requires a FFT with $N' = M_{\text{Tx}}N$ complex points and 2 reconstruction filters (or 4 if $f_c \neq f_0$).

2.3.1.2 Baseband Approach

In this case, the signal associated to a given burst is given by (2.3). Sampling the reference burst (2.18) at rate $N'/T = 2M_{\text{Tx}}N/T$, we obtain the samples

$$s_n^{M_{\text{Tx}}} \triangleq \text{Re}\left\{s_n^{(P)}\right\} = \text{Re}\left\{s^{(P)}(t)\Big|_{t=\frac{nT}{N'}}\right\}$$

$$= \text{Re}\left\{\sum_{k=0}^{N-1} \ddot{S}_k\, e^{j2\pi\frac{kn}{N'}}\right\}, \tag{2.40}$$

$n = 0, 1, \ldots, N' - 1$. Defining s_n as in (2.28), we find that $\{s_n; n = 0, 1, \ldots, N' - 1\} = \frac{1}{N'}\,\text{IDFT}\,\{S_k; k = 0, 1, \ldots, N' - 1\}$, and using

$$S_k = \begin{cases} \tilde{S}_k, & 0 \le k \le N \\ 0, & k = 0 \text{ or } N+1 \le k \le N' - N - 1 \\ \tilde{S}_{N'-k}^*, & N' - N \le k \le N' - 1 \end{cases} \tag{2.41}$$

instead of (2.29) (see Figure 2.13), the resulting time domain samples $s_n^{(P)}$ are real and, as a consequence, samples $s_n^{M_{\text{Tx}}} = \text{Re}\{s_n^{(P)}\} = s_n^{(P)}$ and s_n are also real. The analog signal associated with a given burst is

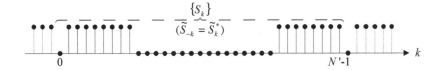

Figure 2.13: Augmented block $\{S_k\}$ given by (2.41) for a baseband DMT transmitter with $N = 8$ and $N' = 4N$.

obtained as described for conventional OFDM, and its complex envelope is given by (2.30). Figure 2.14 illustrates the structure of a baseband DMT transmitter and receiver.

2.3.2 Orthogonal Frequency Division Multiple Access

An Orthogonal Frequency Division Multiple Access (OFDMA) scheme [SLK97, KR02] is a multi-user extension of OFDM that allows multiple access on the same channel, with a different set of subcarriers assigned to each user. Therefore, it combines an OFDM modulation with

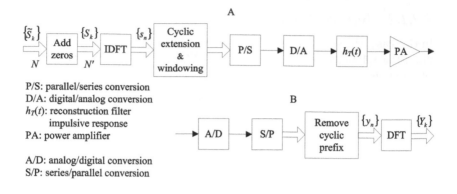

A

P/S: parallel/series conversion
D/A: digital/analog conversion
$h_T(t)$: reconstruction filter
    impulsive response
PA: power amplifier

A/D: analog/digital conversion
S/P: series/parallel conversion

Figure 2.14: Baseband DMT transmitter (A) and receiver (B) structure.

a Frequency Division Multiple Access (FDMA) scheme. Moreover, it is suitable for severe frequency-selective channels and allows a flexible and efficient management of the spectrum. OFDMA distributes subcarriers among users so all users can transmit and receive at the same time within a single channel on what are sometimes called subchannels. The way subcarriers are assigned to a given user can be matched to provide the best performance, which results in fewer problems with fading and interference based on the location and propagation characteristics of each user. It is used in wireless broadband access technologies IEEE 802.16a/d/e, commonly referred to as WiMAX and 3rd Generation Partnership Project (3GPP) Long Term Evolution (LTE). It is also the access method candidate for Wireless Regional Area Networks (WRAN).

Let us consider an OFDMA system with P users and N_p subcarriers assigned to the pth user. The total number of subcarriers is N', the number of in-band subcarriers (i.e., the number of subcarriers that can be assigned to users) is N (it is assumed that $N \geq \sum_{p=1}^{P} N_p$), and we have $N' - N$ subcarriers that are always idle. The idle subcarriers are used to simplify the design of the reconstruction filter; they can also be employed to define unused regions of the spectrum (e.g., when the transmission band is fragmented [DFG$^+$05]).

2.3.2.1 Downlink Transmission

The OFDMA downlink (i.e., transmission from the base station to the mobile terminals) is essentially equivalent to an OFDM system. The difference is that in OFDMA, each transmitted block conveys simultaneous information for multiple users, while in OFDM it carries data for a single specific user. We assume that the base station communicates with P users using the N useful subcarriers. A set Ψ_p is associated with the pth user, containing the indices of the subcarriers assigned to that user. Clearly, to avoid that a given subcarrier is shared by different users, the sets Ψ_p must be mutually exclusive, i.e., $\Psi_p \cap \Psi_{p'} = \emptyset$, for $p \neq p'$. Without loss of generality, we consider the situation in which only one set is assigned to each user, even though in practice more sets may be allocated to the same user depending on its requested data rate. It is usual to refer to set Ψ_p as a subchannel, which is equivalent to consider that the N useful subcarriers are divided into R subchannels, each consisting of N_r subcarriers, with $\sum_{r=1}^{R} N_r = N$. Since the maximum number of users that the system can simultaneously support is limited to R, obviously $R \geq P$.

There are several possible methods to distribute subcarriers among active users (or subchannels), generally designated as Carrier Assignment Schemes (CASs) [MKP07]. In this work, we consider two different CASs: Regular Grid (RG) and Block Assignment (BA) (also known as subband CAS), as shown in Figure 2.15. With block assignment, each subchannel is composed of a group of N_p adjacent subcarriers, as shown in Figure 2.15.B. The major drawback of this approach is that it does not exploit the frequency diversity offered by the multipath channel, since a deep fade might hit a substantial number of subcarriers of a given user. Regular grid provides a viable solution to this problem. It is obtained by adopting the interleaved CAS shown in Figure 2.15.A, where the subcarriers of each user are uniformly spaced over the signal bandwidth at a distance R from each other. RG higher multipath diversity is a consequence of not using adjacent subcarriers for the same user, leading to fewer correlated channel coefficients. However, RG is not suitable for the uplink, since it requires perfect carrier synchronization between users (for BA, we can employ idle subcarriers between blocks to reduce the degradation due to carrier synchronization errors between users). Although this method can fully exploit the channel frequency diversity, the current trend in OFDMA favours a more flexible allocation strategy called generalized CAS, where users can select the best subcarriers (i.e., those with the highest Signal-to-Noise Ratio (SNR)) that are available at the moment. Since there is no rigid association between subcarriers and users, the generalized CAS allows dynamic resource allocation and provides more flexibility than block assignment or regular grid.

Regardless of the adopted CAS, the OFDMA downlink signal can be written as the sum of bursts given by (2.1), with each burst given by (2.30). The frequency-domain block transmitted on the downlink is

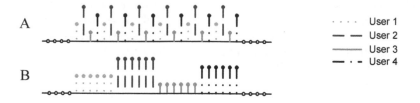

Figure 2.15: Carrier assignment schemes: Regular Grid (A) and Block Assignment (B).

$\{S_k; k = 0, 1, \ldots, N' - 1\}$, with

$$S_k = \begin{cases} \xi_p \tilde{S}_k^{(p)}, & p = 0, 1, \ldots, P; k \in \Psi_p \\ 0, & \text{otherwise}, \end{cases} \tag{2.42}$$

where the set $\{\tilde{S}_k^{(p)}; k \in \Psi_p\}$ associated to the pth user depends on the adopted CAS and ξ_p is an appropriate weighting coefficient that accounts for power control issues.

2.3.2.2 Uplink Transmission

Figure 2.16 illustrates the system in the uplink case (i.e., the transmission from the mobile terminals to the base station). Each user can be seen as an OFDM transmitter, hence the frequency-domain block to be transmitted by the pth user is $\{S_k^{(p)}; k = 0, 1, \ldots, N' - 1\}$, where

$$S_k^{(p)} = \begin{cases} \xi_p \tilde{S}_k^{(p)}, & k \in \Psi_p \\ 0, & \text{otherwise} \end{cases} \qquad p = 0, 1, \ldots, P \tag{2.43}$$

and we have P time-domain signals of the form (2.30), i.e.,

$$s^{(p)}(t) = \sum_{n=-\infty}^{+\infty} s_n^{(p)} w_n h_T \left(t - \frac{nT}{N'} \right), \tag{2.44}$$

where $\{s_n^{(p)}; n = 0, 1, \ldots, N' - 1\} = \frac{1}{N'} \text{IDFT} \{S_k^{(p)}; k = 0, 1, \ldots, N' - 1\}$.

2.3.3 Multicarrier Code Division Multiple Access

Multicarrier Code Division Multiple Access (MC-CDMA) [YLF93,HP97, CNS+05] is a multiple access technique that combines the principle of

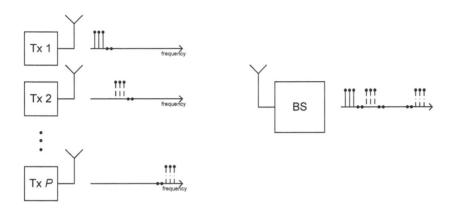

Figure 2.16: OFDMA uplink system with p users.

coded division multiplexing with OFDM modulation, typically employing orthogonal spreading codes. MC-CDMA schemes are used in the 4th generation of high-speed wireless multimedia communications systems.

It aims to provide multiple access capability and mitigate multipath interference. Like OFDM, the MC-CDMA signal is made up of a series of equal amplitude subcarriers. Unlike OFDM, where each subcarrier transmits a different symbol, MC-CDMA transmits every data symbol on multiple narrowband subcarriers. As it is unlikely for all subcarriers to experience deep fade simultaneously, frequency diversity is achieved when the subcarriers are suitably combined at the receiver. Typical combining techniques include Equal Gain Combining (EGC), Maximum Ratio Combining (MRC), Minimum Mean Square Error Combining (MMSEC) and Orthogonality Restoring Combining (ORC). MC-CDMA applies spreading in the frequency domain by mapping a different chip of the spreading sequence to an individual OFDM subcarrier.

CDMA uses unique spreading codes to spread the baseband data before transmission. The signal is transmitted in a channel at a level below noise level. The receiver then uses a correlator to despread the wanted signal, which is passed through a narrow bandpass filter. Unwanted signals will not be despread and will not pass through the filter. Codes take the form of a carefully designed one/zero sequence produced at a much higher rate than that of the baseband data. The rate of a spreading code is referred to as chip rate rather than bit rate. The transmitter and receiver of an OFDMA and MC-CDMA transmission scheme differ only in the subcarrier allocation and the additional spreading and detection component for MC-CDMA.

2.3.3.1 Downlink Transmission

This section considers the use of MC-CDMA schemes in the downlink transmission. Figure 2.17 illustrates the transmission model in the downlink case. The transmitter and receiver of an OFDMA and MC-CDMA transmission scheme differ basically in the subcarrier allocation and the additional spreading and detection component for MC-CDMA.

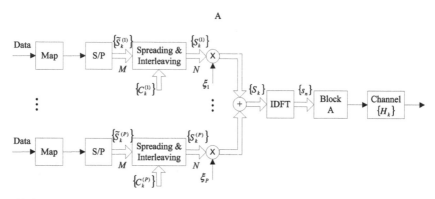

'Block A': add cyclic extension, windowing, P/S conversion and block 'Tx'

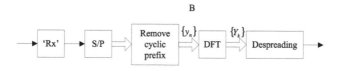

Figure 2.17: MC-CDMA downlink transmitter (A) and receiver (B) structure.

Let P be the number of spreading codes and M be the number of data symbols for each spreading code. A sequence of P blocks of data symbols $\{\tilde{S}_m^{(p)}; m = 0, 1, \ldots, M-1\}$ associated to the pth spreading code are selected from given constellations with an appropriate mapping rule and then serial-to-parallel converted. Each block of data symbols is submitted to a frequency-domain spreading operation by multiplication with a spreading sequence and, to improve diversity gains, different chips associated to a given data symbol should not be transmitted by adjacent subcarriers. Using an orthogonal spreading with K-length Walsh-Hadamard sequences (K is known as spreading factor, spreading gain or processing gain), the kth chip for the pth spreading code is obtained from

$$S_k^{(p)} = C_k^{(p)} \tilde{S}_{k \bmod M}^{(p)}, \qquad k = 0, 1, \ldots, N - 1 \tag{2.45}$$

($x \bmod y$ denotes 'reminder of division of x by y'), where $N = KM$ and $\{C_k^{(p)}; k = 0, 1, \ldots, N - 1\}$ is the spreading sequence associated to the pth spreading code. Generally, an orthogonal spreading is assumed, with $|C_k^{(p)}| = 1$.

The frequency-domain block to be transmitted is $\{S_k; k = 0, 1, \ldots, N - 1\}$, with each symbol obtained from

$$S_k = \sum_{p=1}^{P} \xi_p S_k^{(p)}, \qquad k = 0, 1, \ldots, N - 1, \tag{2.46}$$

where ξ_p is an appropriate weighting coefficient that accounts for the different powers assigned to different spreading codes (the power associated to the pth spreading code is proportional to $|\xi_p|^2$). An IDFT operation generates the sampled version of the time-domain MC-CDMA signal $\{s_n; n = 0, 1, \ldots, N' - 1\}$. A CP is added and time windowing performed, followed by parallel to series conversion and a 'Tx' block identical to the one shown in Figure 2.6.B. The transmitted signal associated to a given data block is given by (2.30). The conventional MC-CDMA receiver is based on a conventional OFDM receiver, as described in Subsection 2.2.2, followed by a despreading procedure.

2.3.3.2 Uplink Transmission

Let us now consider the use of MC-CDMA schemes in the uplink transmission. Since signals associated to different users are affected by different propagation channels, the transmission model is significantly different. Another issue is the loss of orthogonality between spreading codes associated to different users, which leads to severe interference levels and, as a consequence, to significant performance degradation, especially when different powers are assigned to different spreading codes or in case the system is fully loaded. Let us consider the MC-CDMA uplink transmission model in a system employing frequency domain spreading involving P users (mobile transmitters) that transmit independent data blocks, as depicted in Figure 2.18. The M data symbols associated to the pth user are serial-to-parallel converted, and the corresponding block $\{\tilde{S}_m^{(p)}; m = 0, 1, \ldots, M - 1\}$ is submitted to frequency-domain spreading and interleaving operations resulting in the frequency-domain transmitted block $\{S_k^{(p)}; k = 0, 1, \ldots, N - 1\}$, where the kth frequency-domain chip associated to the pth user is given by

$$S_k^{(p)} = \xi_p C_k^{(p)} \tilde{S}_{k \bmod M}^{(p)}, \qquad k = 0, 1, \ldots, N - 1, \tag{2.47}$$

A

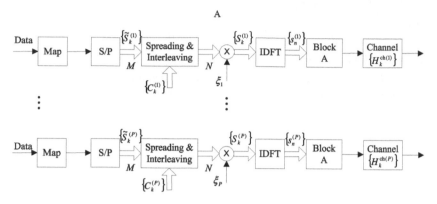

'Block A': add cyclic extension, windowing, P/S conversion and block 'Tx'

B

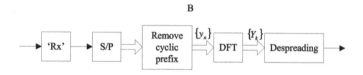

Figure 2.18: MC-CDMA uplink transmitter (A) and receiver (B) structure.

again with $N = KM$, and ξ_p is an appropriate weighting coefficient that accounts for the propagation losses of the pth user. It is usual to assume a pseudo-random spreading (with $C_k^{(p)}$ belonging to a Quadrature Phase-Shift Keying (QPSK) constellation and $|C_k^{(p)}| = 1$), since this type of sequences present better cross-correlation properties when compared with the orthogonal sequences used in an asynchronous environment. As in the downlink case, a CP is inserted before transmission, followed by time windowing, parallel to series conversion and a 'Tx' block.

The signal received at the Base Station (BS) is sampled at the chip rate leading, after discarding the samples associated to the CP, to the time-domain block $\{y_n; n = 0, 1, \ldots, N - 1\}$, which is passed to the frequency-domain, leading to the block $\{Y_k; k = 0, 1, \ldots, N - 1\}$ with

$$Y_k = \sum_{p=1}^{P} \tilde{S}_{k \bmod M}^{(p)} H_k^{(p)} + N_k, \qquad k = 0, 1, \ldots, N - 1, \qquad (2.48)$$

with

$$H_k^{(p)} = \xi_p H_k^{\mathrm{ch}(p)} C_k^{(p)}, \qquad (2.49)$$

which can be regarded as the 'overall' channel frequency response at the kth subcarrier for the pth user, with $H_k^{\text{ch}(p)}$ denoting the channel frequency response at the kth subcarrier for the pth user and N_k denoting the corresponding channel noise. Let us define the length-K column vector containing the received frequency-domain samples

$$\mathbf{Y}_k = \begin{bmatrix} Y_k & Y_{k+M} & \cdots & Y_{k+(K-1)M} \end{bmatrix}^T \tag{2.50}$$

and the length-P column vector

$$\mathbf{S}_k = \begin{bmatrix} \tilde{S}_{k\,\text{mod}\,M}^{(1)} & \tilde{S}_{k\,\text{mod}\,M}^{(2)} & \cdots & \tilde{S}_{k\,\text{mod}\,M}^{(P)} \end{bmatrix}^T \tag{2.51}$$

containing the mth data symbol of each user. Hence, (2.48) can be written in matrix form as

$$\mathbf{Y}_k = \mathbf{H}_k \mathbf{S}_k + \mathbf{N}_k, \tag{2.52}$$

where

$$\mathbf{N}_k = \begin{bmatrix} N_k & N_{k+M} & \cdots & N_{k+(K-1)M} \end{bmatrix}^T \tag{2.53}$$

denotes the length-K column vector containing the noise samples associated to the set of frequencies Ψ_m employed in the transmission of the mth data symbol of each user, and $\mathbf{H}_k$ is the $K \times P$ 'overall' channel frequency response matrix associated to $\mathbf{S}_k$, i.e.,

$$\mathbf{H}_k = \begin{bmatrix} H_k^{(1)} & H_k^{(2)} & \cdots & H_k^{(P)} \\ H_{k+M}^{(1)} & H_{k+M}^{(2)} & \cdots & H_{k+M}^{(P)} \\ \vdots & \vdots & \ddots & \vdots \\ H_{k+(K-1)M}^{(1)} & H_{k+(K-1)M}^{(2)} & \cdots & H_{k+(K-1)M}^{(P)} \end{bmatrix}, \tag{2.54}$$

with columns associated to the different users and lines associated to the set of frequencies Ψ_m.

2.3.4 Software Radio

The main idea behind software radios is to employ a wideband Analog-to-Digital Converter (ADC), and all subsequent processing is implemented digitally [Mit95]. By employing software radio, the cost of the base station can be substantially reduced, since a single transceiver is required, instead of a transceiver for each channel. Moreover, we increase the flexibility of the base stations, which can be programmed for the different existing standards, as well as for undefined future standards [ZK99, TBT99].

Figure 2.19 presents a software radio architecture. We consider P channels, each one with a bandpass signal centered on the frequency f_p,

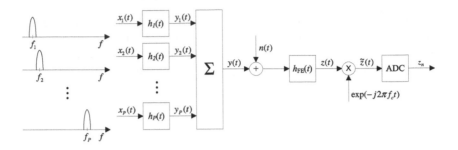

Figure 2.19: Software radio architecture.

$p = 1, 2, \ldots, P$. Without loss of generality, it is assumed that the Power Spectral Densities (PSDs) associated to the different channels do not overlap. The overall signal at the input of the receiver is

$$y(t) = \sum_{p=1}^{P} y_p(t), \tag{2.55}$$

plus the channel noise, where the signal associated to the pth channel is given by

$$y_p(t) = \mathrm{Re}\left\{\tilde{y}_p(t)\, e^{j2\pi f_p t}\right\}, \tag{2.56}$$

with $\tilde{y}_p(t)$ denoting its complex envelope, referred to the frequency f_p. After the front-end filter, with impulse response $h_{\mathrm{FE}}(t)$ and assumed transparent to $y(t)$, we obtain the signal $z(t)$ whose complex envelope referred to the frequency f_c is

$$\tilde{z}(t) = \sum_{p=1}^{P} \tilde{y}_p(t)\, e^{j2\pi(f_p - f_c)t} + n(t), \tag{2.57}$$

with $n(t)$ denoting the noise component. It is assumed that $f_c = f_0 - \Delta f$, with f_0 denoting the central frequency of $y(t)$ and B its bandwidth (see Figure 2.20). This means that the complex envelopes are referred to a frequency that is shifted from the central frequency of $y(t)$ (i.e., the usual reference frequency) by a factor Δf. It should be noted that the selection of $f_c \neq f_0$ (i.e., $\Delta f \neq 0$) implies an asymmetrical front-end filtering prior to the ADC.

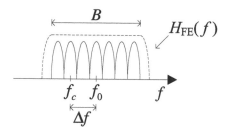

Figure 2.20: Reference frequency for the complex envelope of a software radio signal.

2.4 Characterization of the Transmitted Signals

In this section, the spectral and time domain characterizations of the transmitted multicarrier signals are presented.

2.4.1 Spectral Characterization

Let us consider a data transmission with equal probability and statistical independence in a conventional OFDM system. In this case, one can assume that different burst are uncorrelated, $E[\tilde{S}_k] = 0$ and

$$E[\tilde{S}_k \tilde{S}_{k'}^*] = \begin{cases} E[|\tilde{S}_k|^2], & k = k' \\ 0, & \text{otherwise} \end{cases} \tag{2.58}$$

($E[\cdot]$ denotes 'ensemble average').

The PSD of the transmitted signal is given by

$$G_s(f) = \frac{1}{T_B} E[|S(f)|^2], \tag{2.59}$$

where $S(f)$ represents the Fourier transform of the complex envelope of the multicarrier burst given by (2.30), i.e.,

$$S(f) = \mathcal{F}\{s(t)\} = \mathcal{F}\left\{ \left(\left(\sum_{n=-\infty}^{+\infty} s_n \, \delta\left(t - n\frac{T}{N'}\right) \right) \cdot w(t) \right) * h_T(t) \right\}$$

$$= \frac{1}{T} \left(\sum_{k=0}^{N'-1} S_k \, W^{\text{eq}}\left(f - \frac{k}{T}\right) \right) \cdot H_T(f), \tag{2.60}$$

with S_k and s_n are given by (2.29) and (2.28), respectively, and

$$W^{\text{eq}}(f) = \sum_{l=-\infty}^{+\infty} W\left(f - \frac{lN'}{T}\right), \tag{2.61}$$

since

$$\mathcal{F}\left\{\sum_{n=-\infty}^{+\infty} s_n \, \delta\left(t - n\frac{T}{N'}\right)\right\} = \frac{1}{T} \sum_{k=-\infty}^{+\infty} S_k \, \delta\left(f - \frac{k}{T}\right). \tag{2.62}$$

Replacing $S(f)$ in (2.59) and noting that

$$E[S_k S_{k'}^*] = \begin{cases} G_{S,k}, & k = k' \\ 0, & \text{otherwise}, \end{cases} \tag{2.63}$$

with $G_{S,k} = E[|S_k|^2]$ (also note that $E[|S_k|^2] = NE[|\tilde{S}_k|^2]/N'$), we get

$$G_s(f) = \frac{|H_T(f)|^2}{T^2 T_B} \sum_{k=0}^{N'-1} G_{S,k} \left| W^{\text{eq}}\left(f - \frac{k}{T}\right) \right|^2. \tag{2.64}$$

If $N' \gg 1$ and $M_{\text{Tx}} \geq 2$, the different replicas of $W(f)$ in $W^{\text{eq}}(f)$ are sufficiently apart and the following approximation can be used

$$G_s(f) \approx \frac{|H_T(f)|^2}{T^2 T_B} \sum_{k=0}^{N'-1} G_{S,k} W_k, \tag{2.65}$$

with $W_k = \left| W\left(f - \frac{k}{T}\right) \right|^2$.

Assuming $|H_T(f)|$ is constant in the band occupied by $\sum_{k=0}^{N'-1} G_{S,k} W_k$, then the PSD of the transmitted signal approximately takes the shape of a sum of functions of type $|W(f)|^2$ centered at $\frac{k}{T}$, $k = 0, 1, \ldots, N-1$, i.e.,

$$G_s(f) \propto \sum_{k=0}^{N'-1} G_{S,k} W_k \tag{2.66}$$

($a \propto b$ means 'a is proportional to b').

We should note that the bursts given by (2.30) suffer two filtering effects that jointly shape the spectrum (see (2.65)): the reconstruction filter and the filtering that is inherent to the use of a non-rectangular window $w(t)$.

Ignoring the amplifier effect, the average power of the transmitted signal is given by

$$\overline{P} = \frac{1}{2} \int_{-\infty}^{+\infty} G_s(f) \, df, \tag{2.67}$$

with $G_s(f)$ given by (2.65). Assuming that $M_{Tx} \geq 2$, $N' \gg 1$ and $H_T(f) = 1$ in the useful region of the spectrum (i.e., for $|f| \leq \frac{N}{2T}$), we can write

$$\overline{P} \approx \frac{1}{2T^2 T_B} \sum_{k=0}^{N'-1} G_{S,k} \int_{-\infty}^{+\infty} |W(f)|^2 df$$

$$- \frac{1}{2T^2 T_B} \sum_{k=0}^{N'-1} G_{S,k} \int_{-\infty}^{+\infty} |w(t)|^2 dt. \tag{2.68}$$

If $w(t)$ is rectangular, with unit amplitude and duration T_B, this expression reduces to

$$\overline{P} \approx \frac{1}{2T^2} \sum_{k=0}^{N'-1} G_{S,k}. \tag{2.69}$$

If we also assume that the available power is equally distributed by the N active subcarriers, then $E[|\tilde{S}_k|^2] = 2\sigma_S^2$ and we simply get

$$\overline{P} \approx \frac{N\sigma_S^2}{T^2}. \tag{2.70}$$

2.4.1.1 *Other Multicarrier Schemes*

The spectral characterization of a DMT signal is independent from the approach used for the transmitter structure and hence can be simply obtained from (2.64). The same happens in the OFDMA downlink case, and the extension for MC-CDMA up and downlink cases is straightforward. As for the uplink OFDMA case, we assume all bursts from the P users are uncorrelated, hence $E[s_n^{(p)}] = E[S_k^{(p)}] = 0$ and

$$E[S_k^{(p)} S_{k'}^{(p)*}] = \begin{cases} G_{S,k}^{(p)}, & k = k' \\ 0, & \text{otherwise,} \end{cases} \tag{2.71}$$

hence the precedent spectral characterization can be used, and the PSD of the pth user $G_s^{(p)}(f)$ has the form of (2.64) for $p = 0, 1, \ldots, P$. The PSD of the overall signal received at the base station is given by

$$G_y(f) = \sum_{p=1}^{P} G_s^{(p)}(f) |H^{(p)}(f)|^2 + G_n(f), \tag{2.72}$$

with $H^{(p)}(f)$ denoting the channel impulse response associated to the pth user. Figure 2.21 illustrates this for a system with four users.

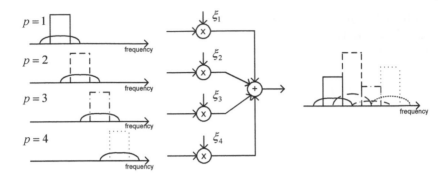

Figure 2.21: PSD for an OFDMA uplink system with four users.

For software radio signals, if we have uncorrelated signals on the different channels, then the PSD of $\tilde{z}(t)$ is

$$G_{\tilde{z}}(f) = \sum_{p=1}^{P} G_{\tilde{y}}^{(p)}(f - f_p + f_c) + G_n(f), \qquad (2.73)$$

with $G_{\tilde{y}}^{(p)}(f) = G_{\tilde{x}}^{(p)}(f)|H^{(p)}(f)|^2$ denoting the PSD of $\tilde{y}_p(t)$, where $H^{(p)}(f)$ denotes the channel impulse response associated to the pth channel, and $G_n(f)$ denoting the PSD of the noise component, which is a straightforward function of $h_{\text{FE}}(t)$ (see Figure 2.19).

2.4.2 Time Domain Characterization

Having in mind the characterization of the time domain samples s_n, let us consider the time domain block $\{s_n;\ n = 0, 1, \ldots, N'-1\} = \text{IDFT} \{S_k;\ k = 0, 1, \ldots, N'-1\}$, with S_k given by (2.29). Again, using the assumption that $E[\tilde{S}_k] = 0$ and $E[\tilde{S}_k \tilde{S}_{k'}^*]$ is given by (2.58), it can be shown that $E[s_n] = 0$ and that the autocorrelation of the time domain samples is given by

$$R_{s,n-n'} = E[s_n s_{n'}^*] = \frac{1}{(N')^2} \sum_{k=0}^{N'-1} \sum_{k'=0}^{N'-1} E[S_k S_{k'}^*]\, e^{j2\pi \frac{kn-k'n'}{N'}}$$

$$= \frac{1}{(N')^2} \sum_{k=0}^{N'-1} G_{S,k}\, e^{j2\pi \frac{k(n-n')}{N'}}, \qquad (2.74)$$

i.e., the block $\{R_{s,n}; n = 0, 1, \ldots, N' - 1\} = \frac{1}{N'}$ IDFT $\{G_{S,k}; k = 0, 1, \ldots, N' - 1\}$. Moreover,

$$R_{s,0} = E[|s_n|^2] = 2\sigma^2 = \frac{1}{(N')^2} \sum_{k=0}^{N-1} G_{S,k}, \qquad (2.75)$$

with σ^2 denoting the variance of the real and imaginary parts of s_n.

If the available power is equally distributed[2] by the N active subcarriers, then $E[|\tilde{S}_k|^2] = 2\sigma_S^2$ and

$$R_{s,n-n'} = \frac{2\sigma_S^2}{(N')^2} \sum_{k=0}^{N-1} e^{j2\pi \frac{k(n-n')}{N'}}$$

$$= 2\sigma^2 \frac{\operatorname{sinc}\left((n - n')N/N'\right)}{\operatorname{sinc}\left((n - n')/N'\right)} e^{-j\pi \frac{n-n'}{N'}}, \qquad (2.76)$$

for $n = 0, 1, \ldots, N' - 1, n' = 0, 1, \ldots, N' - 1$, with $\sigma^2 = \frac{N}{(N')^2}\sigma_S^2$. When the number of used subcarriers is high ($N \gg 1$), the following approximation can be used

$$R_{s,n-n'} \approx 2\sigma^2 \operatorname{sinc}\left(\frac{(n - n')N}{N'}\right) e^{-j\pi \frac{n-n'}{N'}}. \qquad (2.77)$$

Note that when there is no oversampling, i.e., if $N' = N$, samples s_n are uncorrelated, since $\operatorname{sinc}(n - n') = 0$ para $n \neq n'$. In case there is oversampling, the correlation between different samples should be taken into account.

When the number of used subcarriers is high and the transmitted complex symbols are uncorrelated, then the real and imaginary parts of s_n (both with zero mean) can be approximately regarded as samples of a zero-mean complex Gaussian process, due to the central limit theorem [Pap84], even for moderate values of N, e.g., $N = 64$ (see Figure 2.22). Hence, $|s_n|$ can be approximated by a Rayleigh distribution.

2.4.2.1 *Other Multicarrier Schemes*

The time domain characterization for a DMT scheme using the bandpass approach is equal to the one for OFDM scheme, we just have to replace samples S_k in (2.58) and (2.74) by samples $S_k^{\Delta N}$. As for the baseband

---

[2]For OFDM schemes, it is usually assumed that $E[|S_k|^2] = G_{s,k}$ is constant in-band and zero out-of-band. For DMT schemes with loading, $E[|S_k|^2] = G_{s,k}$ is not necessarily constant in-band, and can even be zero for subcarriers at frequency notches.

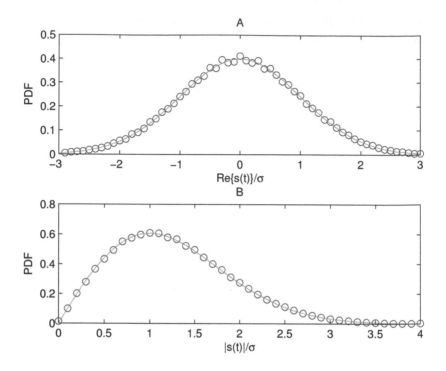

Figure 2.22: Probability density function of Re$\{s(t)\}$ (or Im$\{s(t)\}$) (A) and $|s(t)|$ (B) when $N = 64$ (o) and for a Gaussian signal (solid line).

case, from (2.41) we can write

$$
E[S_k S_{k'}^*] = \begin{cases} E[|S_k|^2], & k' = k \\ E[S_k^2], & k' = N' - k \\ 0, & \text{otherwise,} \end{cases} \tag{2.78}
$$

As before, we assume $E[S_k] = 0$, hence $E[s_n] = 0$ and

$$
R_{s,n-n'} = E[s_n s_{n'}^*] = \frac{1}{(N')^2} \sum_{k=0}^{N'-1} \sum_{k'=0}^{N'-1} E[S_k S_{k'}^*] e^{j2\pi \frac{kn - k'n'}{N'}}
$$

$$
= \frac{1}{(N')^2} \sum_{k=0}^{N'-1} E[|S_k|^2] e^{j2\pi \frac{k(n-n')}{N'}} + \frac{1}{(N')^2} \sum_{k=0}^{N'-1} E[S_k^2] e^{j2\pi \frac{k(n+n')}{N'}}.
$$

$$
\tag{2.79}
$$

For constellations with equiprobable points that satisfy the condition

$$S_k \in S \Rightarrow S_k e^{j\frac{\pi}{2}} \in S, \tag{2.80}$$

where S represents the set of points of the constellation, we have $E[S_k^2] = 0$ (note that the constellations considered in this work, square QAM, cross QAM and Binary Phase-Shift Keying (BPSK) satisfy this condition) and (2.79) reduces to

$$R_{s,n-n'} = \frac{1}{(N')^2} \sum_{k=0}^{N'-1} E[|S_k|^2] e^{j2\pi \frac{k(n-n')}{N'}}. \tag{2.81}$$

Setting $G_{S,k} = E[|S_k|^2]$ yields (2.74).

As for OFDMA, in the downlink case the time domain analysis is equal to the OFDM scheme. Let us now consider the OFDMA uplink case. When the number of used subcarriers is high[3] ($\sum_p N_p \gg 1$), the time-domain coefficients s_n can be approximately regarded as samples of a zero-mean complex Gaussian process, and the autocorrelation of the signal is given by (2.74). The autocorrelation of the pth user takes the form of (2.74) and is given by

$$R_{s,n-n'}^{(p)} = E[s_n^{(p)} s_{n'}^{(p)*}] = \frac{1}{(N')^2} \sum_{k=0}^{N'-1} G_{S,k}^{(p)} e^{j2\pi \frac{k(n-n')}{N'}}, \tag{2.82}$$

where $G_{S,k}^{(p)} = E[|S_k^{(p)}|^2]$ and $\{R_{s,n}^{(p)}; n = 0, 1, \dots, N'-1\} = \frac{1}{N'}$ IDFT $\{G_{S,k}^{(p)}; k = 0, 1, \dots, N'-1\}$, and

$$R_{s,0}^{(p)} = E[|s_n^{(p)}|^2] = 2\sigma_p^2 = \frac{1}{(N')^2} \sum_{k=0}^{N'-1} G_{S,k}^{(p)}, \tag{2.83}$$

with σ_p^2 denoting the variance of the real and imaginary parts of $s_n^{(p)}$. The extension for MC-CDMA up and downlink cases is straightforward.

The autocorrelation function of a software radio signal is given by

$$R_{\tilde{z}}(\tau) = E[\tilde{z}(t)\tilde{z}^*(t-\tau)] = \sum_{p=1}^{P} R_{\tilde{y}}^{(p)}(\tau) e^{j2\pi(f_p - f_c)\tau} + R_n(\tau), \tag{2.84}$$

with $R_{\tilde{y}}^{(p)}(\tau)$ and $R_n(\tau)$ denoting the autocorrelation functions of $\tilde{y}_p(t)$ and $n(t)$, respectively.

---

[3]The number of active subcarriers can be a small fraction of N for OFDMA signals, especially when the number of users is high.

As seen in this section, OFDM schemes and their variants have a Gaussian nature. Due to this, they are prone to nonlinear distortion effects. The aim of this work is to study the impact of these effects on several systems, taking advantage of the mentioned Gaussian nature.

2.4.3 Envelope Fluctuations

When transmitting over a sinusoidal carrier, the power amplification linearity requirements grow with the envelope fluctuations. It is well-known that SC with constant amplitude constellations (e.g., PSK) allows significant reductions of those fluctuations and do not require linear amplification.

For a multicarrier signal, even with constant envelope adopted constellations, the corresponding time domain samples have high envelope fluctuations caused by the large number of independent subcarriers with random phase and amplitude added together at the modulator. Figure 2.23 illustrates this by showing the envelope of a multicarrier signal with $N = 256$ subcarriers and $M_{\mathrm{Tx}} = 4$. As mentioned before, using the

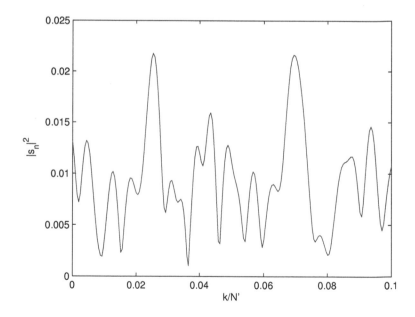

Figure 2.23: Detail of the evolution of the envelope of a multicarrier signal with $N = 256$ subcarriers and $M_{\mathrm{Tx}} = 4$.

central limit theorem [Pap84], if the complex symbols transmitted on the different subcarriers are uncorrelated, then both the real and imaginary parts of the time domain samples s_n are approximately Gaussian, even for moderate values of N. Hence, samples $|s_n|$ are approximately Rayleigh distributed (see Figure 2.22) and consequently the high envelope fluctuations of the multicarrier signal will probably cause power amplification problems.

To characterize the magnitude of the envelope fluctuations, it is usual to use the Peak-to-Mean Envelope Power Ratio (PMEPR), defined as

$$\text{PMEPR} \triangleq \frac{P_{\text{peak}}}{\overline{P}}, \tag{2.85}$$

where $\overline{P}$ denotes the average power of the signal, and P_{peak} denotes the corresponding peak power and is equal to half the maximum envelope power. It is easy to find that

$$P_{\text{peak}} \propto N^2 \max |S_k|^2, \tag{2.86}$$

and that, since from (2.69),

$$\overline{P} \propto \sum_{k=0}^{N'-1} E[|S_k|^2], \tag{2.87}$$

we find that

$$\text{PMEPR} \propto \frac{N^2 \max |S_k|^2}{\sum_{k=0}^{N'-1}, E[|S_k|^2]} \tag{2.88}$$

which means that the PMEPR depends only on the used constellations. If the available power is equally distributed by the N active subcarriers, we can also write (see (2.70))

$$\text{PMEPR} \propto N. \tag{2.89}$$

Clearly, the PMEPR grows with N and can take extremely high values. Since for $N \gg 1$ the highest envelope values have a very small probability (see Figure 2.24), this PMEPR definition can be a pessimistic parameter to evaluate power amplification requirements in multicarrier systems. In this case, it is reasonable to adopt an alternative statistical definition, as follows

$$\text{PMEPR} \triangleq \frac{X^2(P)}{\overline{P}}, \tag{2.90}$$

where $X(P)$ is the envelope value that is exceeded with probability P. Since $s_{\text{MC}}(t)$ is approximately Gaussian distributed with variance σ^2, the corresponding Probability Density Function (PDF) for $R = |s_{\mathrm{MC}}(t)|$ is

$$p(R) = \frac{R}{\sigma^2} e^{-\frac{R^2}{2\sigma^2}}, \qquad (2.91)$$

which means that the probability of the envelope exceeding X is given by

$$P = \mathrm{Prob}(R > X) = \int_X^{+\infty} p(R)dR = e^{-\frac{X^2}{2\sigma^2}} \qquad (2.92)$$

and $X(P) = \sqrt{-2\sigma^2 \log(P)}$ [DG04]. A reasonable value for P is 10^{-3}, and it is clear from Figure 2.24 that it corresponds to PMEPR ≈ 8.4 dB, regardless of the value of N (as long as $N \gg 1$).

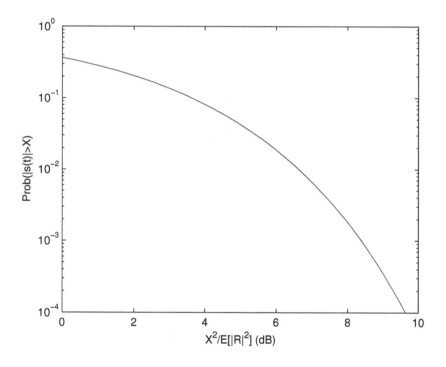

Figure 2.24: Probability of the envelope of a complex Gaussian signal exceeding X.

Chapter 3

Nonlinear Distortion of Gaussian Signals

One of the major drawbacks of multicarrier systems is their vulnerability to nonlinear distortion effects, which come from the fact that most of the components of the transmitter and receiver do not show perfectly linear behaviours. These include the Discrete Fourier Transform (DFT)/ Inverse Discrete Fourier Transform (IDFT) pair, the high-power amplifier and the digital-to-analog and analog-to-digital converters, among others.

The development of analytical techniques to assess the impact of nonlinearities on the performance of multicarrier signals has been a subject of extensive research. However, the fact that such techniques focus on the performance of signals distorted by specific nonlinear devices, such as clipping devices or high-power amplifiers, causes dispersion of techniques and results.

Study of nonlinear effects, as well as performance evaluation of multicarrier systems where they occur, usually resorts to Monte Carlo simulations, which require large computation times. When the number of subcarriers is high, multicarrier signals exhibit a Gaussian-like behaviour, which can be used to analytically characterize a multicarrier signal submitted to a nonlinear device, and this characterization can then be employed for performance evaluation of nonlinearly distorted multicarrier signals [BC00, DTV00, DG04].

In this chapter, it is shown how we can obtain the statistical characterization of nonlinear distorted Gaussian signals. The adopted approach is not original and relies on several analytical results related to the evaluation of nonlinear distortion effects on Gaussian signals dispersed on literature; see, for example, [Ric45, GV94, BC00, Dar03, DG04].

The major outcomes that result from this Gaussian approximation are:

- Following Price's theorem [Pri58], the nonlinearly distorted multicarrier signals can be decomposed in uncorrelated useful and self-interference components;

- Using classical Intermodulation Product (IMP) analysis [Ste74], we can obtain analytically the Power Spectral Density (PSD) of the self-interference component, as well as the PSD of the overall transmitted signal [DG04];

- The signal-to-interference ratio and signal-to-noise plus interference ratio can be obtained analytically [BC00, DTV00, DG04];

- The nonlinear self-interference component is Gaussian at the subcarrier level, allowing an accurate computation of the Bit Error Rate (BER) performance [DG04].

Section 3.1 presents the characterization of the three types of nonlinearities considered in this book. Section 3.2 is dedicated to the impact of memoryless nonlinearities on Gaussian signals. The presented analysis is extended to Cartesian nonlinearities in Section 3.3. The impact of polar memoryless nonlinearities on Gaussian signals is studied in Section 3.4. The results from Sections 3.2–3.4 are analytically particularized in Appendix B for some specific nonlinearities.

Results presented in this chapter will be used in Chapter 4 for the sake of comparison and in Chapter 5 for the performance analysis of several multicarrier systems.

3.1 Memoryless Nonlinearities Characterization

In this book, we consider three types of nonlinearities: memoryless nonlinear devices, Cartesian memoryless nonlinear devices and polar memoryless nonlinear devices.

3.1.1   Memoryless Nonlinear Characteristics

The output of a memoryless nonlinear device depends on the signal at the input and can be written as

$$y(t) = g(x(t)), \tag{3.1}$$

where $g(x)$ is a nonlinear function of x, and $x(t)$ and $y(t)$ are real functions of t and represent, respectively, the input and output signals of the nonlinear device.

We are generally interested in memoryless odd nonlinear characteristics (i.e., $g(-x) = -g(x)$) that can be expanded as a power series of the form

$$g(x) = \sum_{m=0}^{+\infty} \beta_m x^{2m+1}. \tag{3.2}$$

In practice, it is usual to use a polynomial with just a few terms as an approximation. The number of terms needed to characterize a given nonlinearity is usually low and depends on the values of $x(t)$. For sufficiently low values, just a term may be enough, since in that case we can assume that $g(x) \propto x$ (e.g., $g(x) = \beta_0 x$ if $|x| << 1$), which means that $g(x)$ reduces to a linear characteristic.

Ideal Clipping

A common memoryless nonlinear device is the ideal clipping of the signal $x(t)$, which can be characterized by

$$g_{\text{clip}}(x) = \begin{cases} -s_M, & x < -s_M \\ x, & |x| \leq s_M \\ s_M, & x > s_M \end{cases} \tag{3.3}$$

for a given clipping level s_M (see Figure 3.1).

As an example of the approximations that can be used for (3.2), Figure 3.2 shows polynomial approximations with 3, 5 and 7 terms for an ideal clipping, optimizing the minimum square error for $x \in [-3, 3]$. Clearly, an approximation with degree 7 reasonably characterizes $g_{\text{clip}}(x)$, provided that $|x| < 3$.

Quantization

A quantizer that operates on samples x can be regarded as a memoryless nonlinearity with

$$x_{\text{quant}} = g_{\text{quant}}(x), \tag{3.4}$$

where $g_{\text{quant}}(x)$ denotes an appropriate quantization characteristic, which is assumed as an odd function of x.

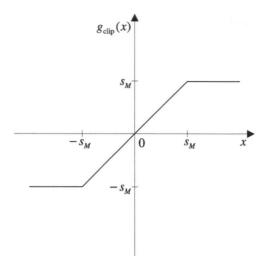

Figure 3.1: Nonlinear characteristic of an ideal clipping.

For example, if we consider floating-point operations, the numerical representation of a given sample x takes the form

$$g_{\text{quant}}(x) = \pm m(x) 2^{-e(x)} s_M, \tag{3.5}$$

where s_M denotes the saturation level (i.e., the clipping effect inherent to the quantization characteristic) and $m(x)$ and $e(x)$ have N_m and N_e bits, respectively (by including the sign bit, the number of bits required to represent x is $N_t = N_e + N_m + 1$). The corresponding binary representations are

$$e(x) = [b_1^{(e)} b_2^{(e)} \cdots b_{N_e}^{(e)}]_2 \tag{3.6}$$

and

$$m(x) = [0.b_1^{(m)} b_2^{(m)} \cdots b_{N_m}^{(m)}]_2, \tag{3.7}$$

where $b_i^{(e)}$ and $b_i^{(m)}$ are the ith bits of $m(x)$ and $e(x)$, respectively ($[\cdot]_2$ indicates binary format). To avoid multiple binary representations of a given x, it is assumed that $[00\cdots0]_2 \leq e(x) \leq [11\cdots1]_2$ (i.e., $0 \leq e(x) \leq 2^{N_e} - 1$) and $[0.00\cdots0]_2 \leq m(x) \leq [0.11\cdots1]_2$, if $e(x) = [11\cdots1]_2$ and $[0.10\cdots0]_2 \leq m(x) \leq [0.11\cdots1]_2$, otherwise (i.e., $0 \leq m(x) \leq 1 - 2^{-N_m}$ if $e(x) = 2^{N_e} - 1$ and $2^{-1} \leq m(x) \leq 1 - 2^{-N_m}$, otherwise). Figure 3.3 presents the evolution of the quantization characteristics, $g_{\text{quant}}(\cdot)$, for $(N_m, N_e) = (5, 0)$ (uniform quantization) and $(N_m, N_e) = (3, 2)$ (non-uniform quantization). It should be noted that, with this representation,

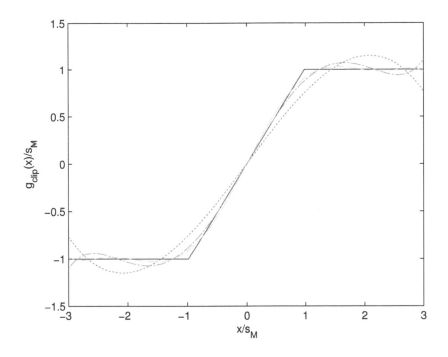

Figure 3.2: Evolution of $g_{\text{clip}}(x)$ for an ideal clipping (solid line) and corresponding polynomial approximations of degree 3 (dotted line), 5 (dash-dotted line) and 7 (dashed line).

the number of quantization levels is 2^{N_t} for $N_e = 0$ and $2^{N_t-1} + 2^{N_m}$ for $N_e \neq 0$.

3.1.2 Cartesian Memoryless Nonlinear Characteristics

Let us now consider a bandpass signal $x_{BP}(t)$. It can be written as

$$x_{BP}(t) = \frac{1}{2}\, x(t)\, e^{j2\pi f_c t} + \frac{1}{2}\, x^*(t)\, e^{-j2\pi f_c t}$$
$$= \text{Re}\left\{ x(t)\, e^{j2\pi f_c t} \right\}, \tag{3.8}$$

where f_c denotes the carrier frequency and $x(t)$ represents the complex envelope of $x_{BP}(t)$ referred to carrier f_c, which can be represented in Cartesian form as

$$x(t) = x_I(t) + j\, x_Q(t), \tag{3.9}$$

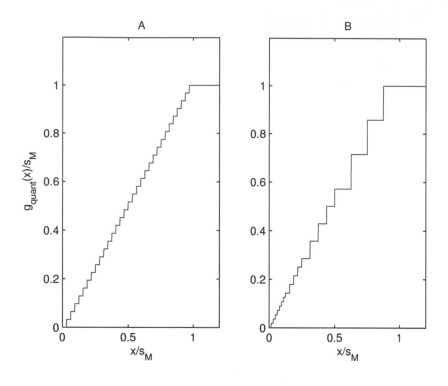

Figure 3.3: Evolution of $g_{\text{quant}}(\cdot)$ for an uniform quantization with $(N_m, N_e) = (5, 0)$ (A) and a non-uniform quantization with $(N_m, N_e) = (3, 2)$ (B).

with $x_I(t) = \text{Re}\{x(t)\}$ and $x_Q(t) = \text{Im}\{x(t)\}$. This representation motivates designating memoryless nonlinear devices that act separately in the real and imaginary parts of a complex signal by Cartesian memoryless nonlinearity (see Section 3.3). Since this operation is performed separately at the in-phase (I) and the quadrature (Q) components of the signal, they are also sometimes designated in literature as I-Q memoryless nonlinearities.

Ideal Cartesian Clipping

The clipping of a complex signal is performed by separately clipping the real and imaginary parts of the signal. This is done in accordance with

$$x_{\text{clip}}(t) = g_{\text{clip}}(x_I(t)) + j\, g_{\text{clip}}(x_Q(t)), \tag{3.10}$$

with $g_{\text{clip}}(x)$ given by (3.3).

Cartesian Quantization

A quantizer that operates on a complex signal $x(t)$ can be regarded as a Cartesian memoryless nonlinearity operating separately at the real and the imaginary parts of the signal. Hence we can write

$$x_{\text{quant}}(t) = g_{\text{quant}}(x_I(t)) + j\, g_{\text{quant}}(x_Q(t)), \tag{3.11}$$

where $g_{\text{quant}}(x)$ is given by (3.5).

3.1.3 Polar Memoryless Nonlinear Characteristics

We will now consider a polar memoryless nonlinear device, i.e., a nonlinear device whose characteristic operates on the amplitude and on the phase of the input signal. This type of nonlinearity is also designated in the literature as a bandpass memoryless nonlinearity [DTV00].

The complex envelope of the bandpass signal given by (3.9) can be written in polar form as

$$x(t) = R(t)\, e^{j\varphi(t)}, \tag{3.12}$$

with $R(t) = |x(t)|$ denoting the complex envelope and $\varphi(t) = \arg(x(t))$ the argument of $x(t)$. Obviously,

$$x_{BP}(t) = \text{Re}\{R(t)\, e^{j\psi(t)}\} = R(t)\cos(\psi(t)), \tag{3.13}$$

with

$$\psi(t) = \varphi(t) + 2\pi f_c t. \tag{3.14}$$

The bandpass signal $R(t)\, e^{j\psi(t)}$, with real part $x_{BP}(t)$, is submitted to the nonlinear device shown in Figure 3.4, where $g_I(\cdot)$ and $g_Q(\cdot)$ are odd real functions. The signal at the output of the nonlinearity is

$$y_P(t) = g_I(R(t)\cos(\psi(t))) - g_Q(R(t)\sin(\psi(t))). \tag{3.15}$$

Since this is a periodic function of $\psi(t)$, it can be expanded in Fourier series, i.e.,

$$y_P(t) = \sum_{l=-\infty}^{+\infty} c_l(R(t))\, e^{jl\psi(t)}, \tag{3.16}$$

with

$$c_l(R) = \frac{1}{2\pi} \int_0^{2\pi} (g_I(R\cos\psi) - g_Q(R\sin\psi))\, e^{-jl\psi}\, d\psi \tag{3.17}$$

(for simplicity, the dependence with t is omitted). This means that

$$\text{Re}\{c_l(R)\} = \frac{1}{2\pi} \int_0^{2\pi} g_I(R\cos\psi)\cos(l\psi)\,d\psi \tag{3.18a}$$

$$\text{Im}\{c_l(R)\} = \frac{1}{2\pi} \int_0^{2\pi} g_Q(R\sin\psi)\sin(l\psi)\,d\psi. \tag{3.18b}$$

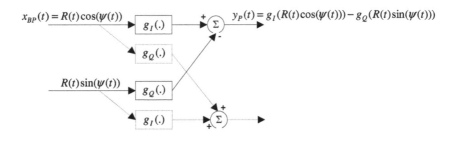

Figure 3.4: Memoryless nonlinearity with bandpass input model.

Since x_{BP} is a bandpass signal, it typically occupies a much smaller band than its carrier frequency f_c; hence, y_P will have several spectral components centered at $\pm l f_c$. If y_P is submitted to a zonal filter transparent to the spectral component centered at f_c and able to totally remove all other components, then the complex envelope of the signal at the filters output can be written as

$$y_{PF}(t) = g(x(t)) = f(R)\,e^{j\varphi(t)}, \tag{3.19}$$

with $g(x(t)) = g(R(t)\,e^{j\varphi(t)}) = f(R)\,e^{j\varphi(t)}$ (i.e., the nonlinear device only operates on the complex envelope of the input signal $x(t)$) and $f(R) = A_I(R) + j\,A_Q(R)$, with

$$A_I(R) = 2\,\text{Re}\{c_1(R)\} = \frac{1}{\pi} \int_0^{2\pi} g_I(R\cos\psi)\cos\psi\,d\psi \tag{3.20}$$

and

$$A_Q(R) = 2\,\text{Im}\{c_1(R)\} = \frac{1}{\pi} \int_0^{2\pi} g_Q(R\sin\psi)\sin\psi\,d\psi. \tag{3.21}$$

Obviously, $f(R) = A(R)\,e^{j\Theta(R)}$, with $A(R) = |f(R)|$ and $\Theta(R) =$

$\arg(f(R))$ denoting the so-called AM-to-AM and AM-to-PM conversions, respectively, i.e., the signal at the output of the polar memoryless nonlinear device can simply be written as

$$y_{PF}(t) = A(R)\, e^{j(\Theta(R)+\varphi)}. \tag{3.22}$$

This means that the effect of the memoryless nonlinearity, represented in Figure 3.4, on $R\,e^{j\psi}$ corresponds, together with the mentioned zonal filter, to the effect on $x_{BP} = \mathrm{Re}\{R\,e^{j\psi}\}$ of a polar memoryless nonlinearity characterized by its AM/AM and AM/PM conversion functions $A(R)$ e $\Theta(R)$, respectively (see Figure 3.5).

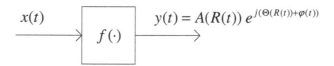

Figure 3.5: Polar memoryless nonlinearity model.

We are interested in polar memoryless nonlinear characteristics $f(R)$ that can be expanded as a power series on R, i.e.,

$$f(R) = A(R)\, e^{j(\Theta(R))} = \sum_{m=0}^{+\infty} \beta_m R^{2m+1}, \tag{3.23}$$

where the coefficients β_m are now complex numbers and, due to the bandpass nature of the nonlinearity, the ones corresponding to even powers are null [BB87]. Taking advantage of this power series, the complex envelope of output signal can be obtained from the complex envelope of the input signal

$$y_{PF}(t) = \sum_{m=0}^{+\infty} \beta_m R^{2m+1}\, e^{j\varphi} = \sum_{m=0}^{+\infty} \beta_m (x(t))^{m+1}(x^*(t))^m. \tag{3.24}$$

As with the memoryless nonlinearities seen in Subsection 3.1.1, most polar memoryless nonlinearities allow polynomial approximations with relatively few terms. In both cases, the output of the nonlinearity can be written as a power series of the input (3.2) for memoryless nonlinearities and (3.24) for polar memoryless nonlinearities. It can be said that each term, except the first, is associated to an IMP, which is responsible for the nonlinear distortion at the output of the nonlinearity [Ric45, Shi71, Ste74].

Ideal Envelope Clipping

A well-known example of a polar memoryless nonlinearity is the ideal envelope clipping, which can be modeled by the function

$$f(R) = A(R) \, e^{j\Theta(R)} = \begin{cases} R, & R \leq s_M \\ s_M, & R > s_M, \end{cases} \tag{3.25}$$

i.e., the AM/PM conversion characteristic is null and the AM/AM conversion characteristic is represented in Figure 3.6, where s_M is the saturation level.

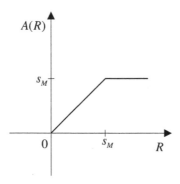

Figure 3.6: AM/AM characteristic of an ideal envelope clipping.

Power Amplifiers

Nonlinear power amplifiers can be modeled as polar memoryless nonlinearities using a Rapp characteristic [Rap91]. As an example, we present the model for a Solid State Power Amplifier (SSPA), which includes envelope clipping as a particular case. Its AM/AM and AM/PM conversion functions are, respectively,

$$A(R) = A_M \frac{R/s_M}{\sqrt[2q]{1 + (R/s_M)^{2q}}} \tag{3.26}$$

and

$$\Theta(R) \approx 0, \qquad (3.27)$$

where A_M is the saturating amplitude (i.e., $\lim_{R\to+\infty} A(R) = A_M$), A_M/s_M is the small signal gain (i.e., $\lim_{R\to 0} A(R)/R = A_M/s_M$), and parameter q is an integer that controls the smooth transition from linear region to saturation region. Figure 3.7 shows the evolution of the AM/AM conversion characteristic for several values of q. Note that, as q gets larger, the AM/AM curve approaches the ideal envelope clipping.

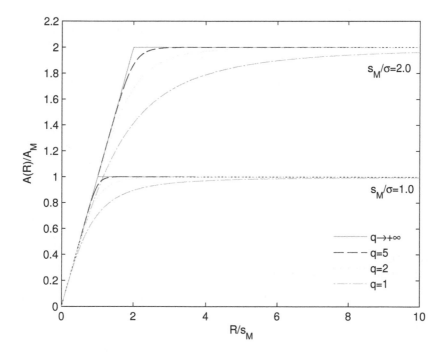

Figure 3.7: Solid State Power Amplifier (SSPA) AM/AM conversion characteristic.

3.2 Memoryless Nonlinear Distortion of Gaussian Signals

In this section, a mathematical analysis that describes the effects of nonlinear distortion on Gaussian signals is presented. This formulation exploits the fact that multicarrier signals with large number of subcarriers are complex Gaussian distributed to characterize the signal at the output of a memoryless nonlinear device.

The well-known Bussgang theorem [Bus52] states that if we have two Gaussian signals $x_1(t)$ and $x_2(t)$ and one of them suffers from nonlinear distortion, then the crosscorrelation function taken after distortion $R_{x_1 y_2}(\tau)$ is identical, except for a factor of proportionality α, to the cross-correlation function taken before the distortion $R_{x_1 x_2}(\tau)$, i.e.,

$$R_{x_1 y_2}(\tau) = \alpha R_{x_1 x_2}(\tau). \tag{3.28}$$

Let us consider a Gaussian signal $x(t)$ with zero mean and variance σ^2 is submitted to a memoryless nonlinearity device characterized by the nonlinear function $g(x)$ (see (3.1)). The autocorrelation of the signal at the input of the nonlinearity is $R_x(\tau) = R_{xx}(\tau) = E[x(t)x(t-\tau)]$, and the cross-correlation between the input and output signals of the nonlinearity, $x(t)$ and $y(t) = g(x(t))$, respectively, is $R_{xy}(\tau) = E[x(t)y(t-\tau)]$. From (3.28) it follows that $R_{xy}(\tau)$ is proportional to the autocorrelation of the input signal, that is

$$R_{xy}(\tau) = \alpha R_x(\tau). \tag{3.29}$$

Using the Bussgang theorem, it can also be shown [Row82, Pap84] that the signal at the output of the nonlinear device can be decomposed into useful and self-interference components as

$$y(t) = \alpha x(t) + d(t), \tag{3.30}$$

with

$$\alpha = \frac{R_{xy}(\tau)}{R_x(\tau)} = \frac{E[xg(x)]}{E[x^2]} = \frac{1}{\sqrt{2\pi}\sigma^3} \int_{-\infty}^{+\infty} xg(x)\, e^{-\frac{x^2}{2\sigma^2}}\, dx \tag{3.31}$$

(for the sake of simplicity, we omit the dependency on t), since the probability density function of the signal $x(t)$ is

$$p(x) = \frac{1}{\sqrt{2\pi}\sigma} e^{-\frac{x^2}{2\sigma^2}}. \tag{3.32}$$

The distortion term $d(t)$ is uncorrelated with the input signal $x(t)$, since

$$R_{xd}(\tau) = E[x(t)d(t-\tau)] = E[x(t)(y(t-\tau) - \alpha x(t-\tau))]$$
$$= R_{xy}(\tau) - \alpha R_x(\tau) = 0. \tag{3.33}$$

The average power of the signal at the nonlinearity output is given by

$$P_{\text{out}} = E[g^2(x)] = \frac{1}{\sqrt{2\pi}\sigma} \int_{-\infty}^{+\infty} g^2(x) e^{-\frac{x^2}{2\sigma^2}} dx, \qquad (3.34)$$

the average power of the useful component is

$$S = |\alpha^2|\sigma^2, \qquad (3.35)$$

and the average power of the self-interference component is simply given by

$$I = P_{\text{out}} - S. \qquad (3.36)$$

Having in mind the statistical characterization of the signal at the output of the nonlinearity, usually $g(x)$ is written in the form of a functional series

$$g(x) = \sum_{m=0}^{+\infty} \beta_m f_m(x), \qquad (3.37)$$

where $\{f_m(x); m = 0, 1, \ldots\}$ is a set of appropriate functions and $\beta_m \in \mathbb{R}$, $m = 0, 1, \ldots$. We are interested in a set of orthogonal polynomial functions in $]-\infty, +\infty[$ relatively to the probability density function of the signal, i.e.,

$$\int_{-\infty}^{+\infty} p(x) f_m(x) f_{m'}(x) dx = 0, \qquad m \neq m', \qquad (3.38)$$

where $p(x)$ is given by (3.32). We will consider the set $\left\{ H_m \left(\frac{x}{\sqrt{2}\sigma} \right); m = 0, 1, \ldots \right\}$, where

$$H_m(x) = (-1)^m e^{x^2} \frac{d^m}{dx^m} \left(e^{-x^2} \right) \qquad (3.39)$$

denotes a Hermite polynomial of degree m [AS72]. It can be shown that

$$\int_{-\infty}^{+\infty} p(x) H_m \left(\frac{x}{\sqrt{2}\sigma} \right) H_{m'} \left(\frac{x}{\sqrt{2}\sigma} \right) dx = \begin{cases} m! 2^m, & m = m' \\ 0, & m \neq m', \end{cases} \qquad (3.40)$$

hence this set of polynomials satisfies (3.38).

The autocorrelation of the signal at the output of the nonlinearity $y(t)$ is

$$R_y(\tau) = E[g(x_1)g(x_2)] = \int_{-\infty}^{+\infty} \int_{-\infty}^{+\infty} g(x_1)g(x_2)p(x_1, x_2)dx_1 dx_2, \qquad (3.41)$$

where $x_1 \triangleq x(t)$ and $x_2 \triangleq x(t-\tau)$ are jointly Gaussian random variables (both with zero mean and variance σ^2), and

$$p(x_1, x_2) = \frac{1}{2\pi\sigma^2\sqrt{1-\rho^2}}\, e^{-\frac{x_1^2+x_2^2-2\rho x_1 x_2}{2\sigma^2(1-\rho^2)}} \tag{3.42}$$

denotes their joint probability density function [Ric45]. The cross-correlation coefficient ρ for the two random variables is the normalized autocorrelation of the signal at the input of the nonlinearity and is given by

$$\rho \triangleq \rho(\tau) = \frac{R_x(\tau)}{R_x(0)}, \tag{3.43}$$

with $R_x(\tau) = E[x(t)x(t-\tau)] = E[x_1 x_2]$ and $R_x(0) = E[x^2(t)] = \sigma^2$.

Direct computation of (3.41) is difficult, since it requires the evaluation of a double integral for each value of τ. However, this computation can be simplified. Let us consider the joint characteristic function of x_1 e x_2 [Pap84], which is given by

$$\Phi(f_1, f_2) = E[e^{j2\pi f_1 x_1} e^{j2\pi f_2 x_2}] = e^{-\frac{(2\pi\sigma)^2}{2}\left(f_1^2+f_2^2+2\rho f_1 f_2\right)}. \tag{3.44}$$

The function $\Phi(f_1, f_2)$ can be expanded as a Taylor series on the variable ρ, yielding

$$\Phi(f_1, f_2) = e^{-\frac{(2\pi\sigma)^2 f_1^2}{2}} e^{-\frac{(2\pi\sigma)^2 f_2^2}{2}} \sum_{m=0}^{+\infty} \left((2\pi\sigma)^2 f_1 f_2\right)^m \frac{\rho^m}{m!}$$

$$= \sum_{m=0}^{+\infty} (2\pi\sigma f_1)^m\, e^{-\frac{(2\pi\sigma)^2 f_1^2}{2}} (2\pi\sigma f_2)^m\, e^{-\frac{(2\pi\sigma)^2 f_2^2}{2}} \frac{\rho^m}{m!}. \tag{3.45}$$

Taking the two-dimensional Fourier Transform of (3.45) we get

$$p(x_1, x_2) = \int_{-\infty}^{+\infty}\int_{-\infty}^{+\infty} \Phi(f_1, f_2)\, e^{j2\pi f_1 x_1 + j2\pi f_2 x_2}\, df_1 df_2$$

$$= \frac{1}{2\pi\sigma^2} \sum_{m=0}^{+\infty} \frac{\sigma^{2m}\rho^m}{m!} \frac{d^n}{dx_1^m}\left(e^{-\frac{x_1^2}{2\sigma^2}}\right) \frac{d^m}{dx_2^m}\left(e^{-\frac{x_2^2}{2\sigma^2}}\right), \tag{3.46}$$

From the definition of Hermite polynomials (see (3.39)), we get

$$\frac{d^m}{dx^m}\left(e^{-\frac{x^2}{2\sigma^2}}\right) = \frac{(-1)^m}{(\sqrt{2}\sigma)^m} H_m\left(\frac{x}{\sqrt{2}\sigma}\right) e^{-\frac{x^2}{2\sigma^2}}, \tag{3.47}$$

hence

$$p(x_1, x_2) = p(x_1)p(x_2) \sum_{m=0}^{+\infty} \frac{\rho^m}{2^m m!} H_m\left(\frac{x_1}{\sqrt{2}\sigma}\right) H_m\left(\frac{x_2}{\sqrt{2}\sigma}\right), \tag{3.48}$$

which is known as Mehler's formula [Sze75]. By using (3.48) in (3.41), we get

$$R_y(\tau) = \int_{-\infty}^{+\infty} \int_{-\infty}^{+\infty} g(x_1)g(x_2)p(x_1)p(x_2)$$

$$\cdot \sum_{m=0}^{+\infty} \frac{\rho^m}{2^m m!} H_m\left(\frac{x_1}{\sqrt{2}\sigma}\right) H_m\left(\frac{x_2}{\sqrt{2}\sigma}\right) dx_1 dx_2$$

$$= \sum_{m=0}^{+\infty} \frac{\rho^m}{2^m m!} \left(\int_{-\infty}^{+\infty} g(x) H_m\left(\frac{x}{\sqrt{2}\sigma}\right) p(x) dx\right)^2. \tag{3.49}$$

Since $g(x)$ is an odd function and Hermite polynomials are even for m even and odd for m odd, we can write

$$R_y(\tau) = \sum_{\gamma=0}^{+\infty} \frac{\rho^{2\gamma+1}}{2^{2\gamma+1}(2\gamma+1)!} \left(\int_{-\infty}^{+\infty} g(x)p(x) H_{2\gamma+1}\left(\frac{x}{\sqrt{2}\sigma}\right) dx\right)^2$$

$$= \sum_{\gamma=0}^{+\infty} P_{2\gamma+1}\rho^{2\gamma+1} = \sum_{\gamma=0}^{+\infty} P_{2\gamma+1}\left(\frac{R_x(\tau)}{R_x(0)}\right)^{2\gamma+1}, \tag{3.50}$$

where $P_{2\gamma+1}$ denotes the total power associated to the IMP of order $2\gamma+1$, given by

$$P_{2\gamma+1} = \frac{1}{2^{2\gamma+1}(2\gamma+1)!} \left(\int_{-\infty}^{+\infty} g(x)p(x) H_{2\gamma+1}\left(\frac{x}{\sqrt{2}\sigma}\right) dx\right)^2. \tag{3.51}$$

Note that, since $H_1(x) = 2x$, the average power of the useful component is $P_1 = |\alpha|^2\sigma^2$, an expected result from (3.30). We can also note the expansion coefficients of $g(x)$ as a Hermite polynomial series in (3.37) are given by

$$\beta_{2\gamma+1} = \sqrt{\frac{P_{2\gamma+1}}{2^{2\gamma+1}(2\gamma+1)!}}. \tag{3.52}$$

Since the useful and self-interference components of $y(t)$ are uncorrelated, the corresponding autocorrelation can be written as

$$R_y(\tau) = |\alpha|^2 R_x(\tau) + R_d(\tau), \tag{3.53}$$

where

$$R_d(\tau) = E[d(t)d(t-\tau)] = \sum_{\gamma=1}^{+\infty} P_{2\gamma+1}\left(\frac{R_x(\tau)}{R_x(0)}\right)^{2\gamma+1} \tag{3.54}$$

is the autocorrelation of the self-interference component, and the average power of the self-interference component is

$$I = P_{\text{out}} - S = R_d(0) = \sum_{\gamma=1}^{+\infty} P_{2\gamma+1}. \tag{3.55}$$

The PSD of $y(t)$ is simply given by

$$G_y(f) = \mathcal{F}\{R_y(\tau)\} = \sum_{\gamma=0}^{+\infty} P_{2\gamma+1}\mathcal{F}\left\{ \left(\frac{R_x(\tau)}{R_x(0)}\right)^{2\gamma+1} \right\}$$

$$= \sum_{\gamma=0}^{+\infty} \frac{P_{2\gamma+1}}{(R_x(0))^{2\gamma+1}} \underbrace{G_x(f) * G_x(f) * \ldots * G_x(f)}_{2\gamma+1}, \tag{3.56}$$

where $G_x(f) = \mathcal{F}\{R_x(\tau)\}$ denotes the PSD of $x(t)$. Clearly,

$$G_y(f) = |\alpha|^2 G_x(f) + G_d(f) \tag{3.57}$$

where $G_d(f) = \mathcal{F}\{R_d(\tau)\}$ denotes the PSD of the self-interference component $d(t)$ and is given by

$$G_d(f) = \mathcal{F}\{R_d(\tau)\} = \sum_{\gamma=1}^{+\infty} P_{2\gamma+1}\mathcal{F}\left\{ \left(\frac{R_x(\tau)}{R_x(0)}\right)^{2\gamma+1} \right\}$$

$$= \sum_{\gamma=1}^{+\infty} \frac{P_{2\gamma+1}}{(R_x(0))^{2\gamma+1}} \underbrace{G_x(f) * G_x(f) * \ldots * G_x(f)}_{2\gamma+1}. \tag{3.58}$$

From (3.56) and (3.58), it is clear that the nonlinearity causes a spectral spreading of the signals, since if $x(t)$ has bandwidth B, then the term associated with the IMP of order $2\gamma + 1$ has bandwidth $(2\gamma + 1)B$.

3.3 Cartesian Memoryless Nonlinearities with Gaussian Inputs

We will now consider a specific situation of bandpass transmission that occurs when the complex envelope of the signal at the output of the nonlinearity $y_C(t)$ is obtained by submitting the real and imaginary parts (i.e., the 'in-phase' and 'quadrature' components) of the signal at the input of the nonlinearity $x_C(t)$, separately, to two identical memoryless nonlinearities that can both be characterized by (3.1). This device can be represented as depicted in Figure 3.8 and will be designated throughout this book as Cartesian memoryless nonlinearity. Next we will apply the results from the preceding section to its characterization.

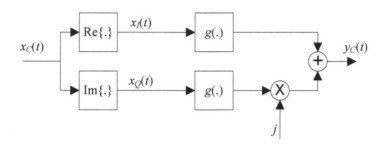

Figure 3.8: Cartesian memoryless nonlinearity model.

The input complex envelope, $x_C(t)$, is modeled as a zero-mean, noise-like, Gaussian process with real and imaginary parts with variance σ^2. Using the Bussgang theorem again, it can be shown that

$$y_C(t) = \alpha_C x_C(t) + d_C(t), \qquad (3.59)$$

where $d_C(t)$ and $x_C(t)$ are uncorrelated and α_C can be obtained from (3.31), with $\mathrm{Re}\{x_C(t)\}$ or $\mathrm{Im}\{x_C(t)\}$ replacing $x(t)$ (note that $\mathrm{Re}\{x_C(t)\}$ and $\mathrm{Im}\{x_C(t)\}$ are Gaussian processes submitted to identical nonlinearities).

To calculate the output autocorrelation, we will proceed in the following way. Clearly, the autocorrelation of $x_C(t)$ is given by

$$\begin{aligned} R_x^C(\tau) &= E[x_C(t)x_C^*(t-\tau)] \\ &= R_{II}(\tau) + R_{QQ}(\tau) + j(R_{QI}(\tau) - R_{IQ}(\tau)), \qquad (3.60) \end{aligned}$$

with

$$R_{II}(\tau) \triangleq E[x_I(t)x_I(t-\tau)] \qquad (3.61\mathrm{a})$$

$$R_{QQ}(\tau) \triangleq E[x_Q(t)x_Q(t-\tau)] = R_{II}(\tau) \qquad (3.61\mathrm{b})$$

$$R_{QI}(\tau) \triangleq E[x_Q(t)x_I(t-\tau)] \qquad (3.61\mathrm{c})$$

$$R_{IQ}(\tau) \triangleq E[x_I(t)x_Q(t-\tau)] = -R_{QI}(\tau), \qquad (3.61\mathrm{d})$$

where $x_I(t) = \text{Re}\{x_C(t)\}$ and $x_Q(t) = \text{Im}\{x_C(t)\}$. This means that

$$R_x^C(\tau) = 2R_{II}(\tau) + j\,2R_{QI}(\tau) \tag{3.62}$$

and, since $R_{II}(0) = R_{QQ}(0) = \sigma^2$ and $R_{IQ}(0) = R_{QI}(0) = 0$, we have $R_x^C(0) = 2\sigma^2$.

The output complex envelope is given by

$$y_C(t) = y_I(t) + jy_Q(t) = g(x_I(t)) + jg(x_Q(t)), \tag{3.63}$$

and its autocorrelation can be written as

$$\begin{aligned}
R_y^C(t) &= E[y_C(t)y_C^*(t-\tau)] \\
&= E[g(x_I(t))g(x_I(t-\tau))] + E[g(x_Q(t))g(x_Q(t-\tau))] \\
&\quad + j(E[g(x_Q(t))g(x_I(t-\tau))] - E[g(x_I(t))g(x_Q((t-\tau))]]).
\end{aligned} \tag{3.64}$$

Note that the four expected values of the last equality of (3.64) involve two jointly Gaussian random variables, submitted to two identical nonlinearities $g(x)$. This means that both have the form (3.41), and consequently, can be obtained from (3.50), with the coefficients $P_{2\gamma+1}$ calculated as in (3.51). Since the coefficients $P_{2\gamma+1}$ are identical for the four expectations of (3.64), we get

$$R_y^C(\tau) = \sum_{\gamma=0}^{+\infty} P_{2\gamma+1}(\rho_{II}^{2\gamma+1} + \rho_{QQ}^{2\gamma+1} + j(\rho_{QI}^{2\gamma+1} - \rho_{IQ}^{2\gamma+1})), \tag{3.65}$$

with

$$\rho_{II} \triangleq \frac{E[x_I(t)x_I(t-\tau)]}{E[|x_I(t)|^2]} = \frac{\text{Re}\{R_x^C(\tau)\}}{R_x^C(0)} \tag{3.66a}$$

$$\rho_{QQ} \triangleq \frac{E[x_Q(t)x_Q(t-\tau)]}{E[|x_Q(t)|^2]} = \frac{\text{Re}\{R_x^C(\tau)\}}{R_x^C(0)} = \rho_{II} \tag{3.66b}$$

$$\rho_{QI} \triangleq \frac{E[x_Q(t)x_I(t-\tau)]}{\sqrt{E[|x_I(t)|^2]E[|x_Q(t)|^2]}} = \frac{\text{Im}\{R_x^C(\tau)\}}{R_x^C(0)} \tag{3.66c}$$

$$\rho_{IQ} \triangleq \frac{E[x_I(t)x_Q(t-\tau)]}{\sqrt{E[|x_I(t)|^2]E[|x_Q(t)|^2]}} = \frac{\text{Im}\{R_x^C(\tau)\}}{R_x^C(0)} = -\rho_{QI}. \tag{3.66d}$$

This means that the autocorrelation of the output of a Cartesian nonlinearity with a bandpass Gaussian input is given by

$$R_y^C(\tau) = 2\sum_{\gamma=0}^{+\infty} P_{2\gamma+1} \frac{(\text{Re}\{R_x^C(\tau)\})^{2\gamma+1} + j(\text{Im}\{R_x^C(\tau)\})^{2\gamma+1}}{(R_x^C(0))^{2\gamma+1}}. \tag{3.67}$$

It should be noted that when the input autocorrelation is real, then (3.67) reduces to equation (9) of [Dar03]. It can also be noted that if $R_x^C(\tau)$ has Hermitian symmetry (i.e., $R_x^C(-\tau) = R_x^{C*}(\tau)$), then $R_y^C(\tau)$ also has Hermitian symmetry.

Equation (3.67) can be rewritten as

$$R_y^C(\tau) = 2 \sum_{\gamma=0}^{+\infty} \frac{P_{2\gamma+1}}{(2R_x^C(0))^{2\gamma+1}} \left[(R_x^C(\tau) + R_x^{C*}(\tau))^{2\gamma+1} \right.$$
$$\left. + (-1)^\gamma (R_x^C(\tau) - R_x^{C*}(\tau))^{2\gamma+1} \right], \quad (3.68)$$

hence the PSD of $y_C(t)$ is simply given by

$$G_y^C(f) = \mathcal{F}\left\{ R_y^C(\tau) \right\}$$
$$= 2 \sum_{\gamma=0}^{+\infty} \frac{P_{2\gamma+1}}{(2R_x^C(0))^{2\gamma+1}} \left[\mathcal{F}\left\{ (R_x^C(\tau) + R_x^{C*}(\tau))^{2\gamma+1} \right\} \right.$$
$$\left. + (-1)^\gamma \mathcal{F}\left\{ (R_x^C(\tau) - R_x^{C*}(\tau))^{2\gamma+1} \right\} \right]$$
$$= 2 \sum_{\gamma=0}^{+\infty} \frac{P_{2\gamma+1}}{(2R_x^C(0))^{2\gamma+1}} f_{2\gamma+1}^C(G_x^C(f)), \quad (3.69)$$

where $G_x^C(f) = \mathcal{F}\{R_x^C(\tau)\}$ denotes the PSD of $x_C(t)$ and

$$f_{2\gamma+1}^C(G(f)) \triangleq \underbrace{(G(f) + G(-f)) * \ldots * (G(f) + G(-f))}_{2\gamma+1}$$
$$+ (-1)^\gamma \underbrace{(G(f) - G(-f)) * \ldots * (G(f) - G(-f))}_{2\gamma+1} \quad (3.70)$$

(note that, since $G_x^C(f)$ is a real, non-negative function of f, then $\mathcal{F}\{R_x^{C*}(\tau)\} = G_x^{C*}(-f) = G_x^C(-f)$). Again, we can write

$$G_y^C(f) = |\alpha_C|^2 G_x^C(f) + G_d^C(f), \quad (3.71)$$

where $G_d^C(f)$ denotes the PSD of the self-interference term $d_C(t)$, which is given by

$$G_d^C(f) = \mathcal{F}\left\{ R_d^C(\tau) \right\} = 2 \sum_{\gamma=1}^{+\infty} \frac{P_{2\gamma+1}}{(2R_x^C(0))^{2\gamma+1}} f_{2\gamma+1}^C(G_x^C(f)). \quad (3.72)$$

3.4 Polar Memoryless Nonlinearities with Gaussian Inputs

Let us consider a bandpass Gaussian signal with real part $x_{BP}(t)$ given by (3.8), whose complex envelope $x(t)$ has zero mean and autocorrelation $R_x^P(\tau) = E[x(t)x^*(t - \tau)]$. If this signal is submitted to a polar memoryless nonlinearity, then the complex envelope of the signal at the output of the nonlinearity can be written as (3.22). If the nonlinearity is characterized by the functions $A(R)$ and $\Theta(R)$, as seen in Section 3.1, this can be rewritten as

$$y_P(t) = A(R) \, e^{j\Theta(R)} \, e^{j\varphi(t)}, \tag{3.73}$$

with $R = R(t) = |x(t)|$, $\varphi = \varphi(t) = \arg(x(t))$, and $A(R)$ and $\Theta(R)$ denote the so-called AM-to-AM and AM-to-PM conversions, respectively.

In Appendix A, it is shown that the autocorrelation of $y_P(t)$ is given by (A.28)

$$R_y^P(\tau) = 2 \sum_{\gamma=0}^{+\infty} P_{2\gamma+1} f_{2\gamma+1}^R(R_x^P(\tau)), \tag{3.74}$$

where

$$f_{2\gamma+1}^R(R(\tau)) \triangleq \frac{(R(\tau))^{\gamma+1} (R^*(\tau))^\gamma}{(R(0))^{2\gamma+1}} \tag{3.75}$$

and $P_{2\gamma+1}$ again denotes the total power associated to the IMP of order $2\gamma + 1$, which in this case can be obtained from (A.26)

$$P_{2\gamma+1} = \frac{1}{4\sigma^6(\gamma+1)} \left| \int_0^{+\infty} R^2 f(R) \, e^{-\frac{R^2}{2\sigma^2}} L_\gamma^{(1)}\left(\frac{R^2}{2\sigma^2}\right) dR \right|^2. \tag{3.76}$$

The PSD of $y_P(t)$ can easily be obtained from (3.74), as follows

$$G_y^P(f) = \mathcal{F}\{R_y^P(\tau)\} = 2 \sum_{\gamma=0}^{+\infty} \frac{P_{2\gamma+1}}{(R_x(0))^{2\gamma+1}} f_{2\gamma+1}^G(G_x^P(f)), \tag{3.77}$$

where $G_x^P(f) = \mathcal{F}\{R_x^P(\tau)\}$ denotes the PSD of $x(t)$ and

$$f_{2\gamma+1}^G(G(f)) \triangleq \underbrace{G(f) * \ldots * G(f)}_{\gamma+1} * \underbrace{G(-f) * \ldots * G(-f)}_{\gamma}. \tag{3.78}$$

From (3.76), we find that $P_1 = |\alpha_P|^2 \sigma^2$, with α_P given by

$$\alpha_P = \frac{1}{2\sigma^4} \int_0^{+\infty} R^2 f(R) \, e^{-\frac{R^2}{2\sigma^2}} dR$$

$$= \frac{1}{2\sigma^4} \int_0^{+\infty} R^2 A(R) \, e^{j\Theta(R)} \, e^{-\frac{R^2}{2\sigma^2}} dR \tag{3.79}$$

and, since the complex envelope $R = |x(t)|$ has Rayleigh distribution, we can write

$$\alpha_P = \frac{E[y_P(t)x^*(t)]}{E[|x(t)|^2]} = \frac{E[RA(R) \, e^{j\Theta(R)}]}{E[R^2]} = \frac{E[Rf(R)]}{E[R^2]}. \tag{3.80}$$

This means that the first IMP is proportional to the input signal with scaling factor α_P. Since the signal corresponding to the other IMPs is uncorrelated with the input signal, then $y_P(t)$ can, as in (3.30), be decomposed into uncorrelated useful and self-interference components

$$y_P(t) = \alpha_P x(t) + d_P(t), \tag{3.81}$$

where $E[y_P(t)x(t-\tau)^*] = 0$ and $d_P(t)$ represents the complex envelope of a self-interference term, which is uncorrelated with $x(t)$. This is an expected result due to the Bussgang theorem, since the input signal is Gaussian.

Since the useful and the self-interference components at the output of the nonlinearity are uncorrelated, we can write

$$R_y^P(\tau) = |\alpha_P|^2 R_x^P(\tau) + R_d^P(\tau), \tag{3.82}$$

where $R_d^P(\tau)$ denotes the autocorrelation of the self-interference component, which is given by

$$R_d^P(\tau) = E[d^P(t)d^{P*}(t-\tau)] = 2\sum_{\gamma=1}^{+\infty} P_{2\gamma+1} f_{2\gamma+1}^R(R_x^P(\tau)). \tag{3.83}$$

The average power of the self-interference component can be obtained from (3.34), with

$$P_{\text{out}} = \frac{1}{2\sigma^2}E[f^2(R)] = \frac{1}{2}\int_0^{+\infty} Rf^2(R)\,e^{-\frac{R^2}{2\sigma^2}}\,dR. \tag{3.84}$$

Note that the average power of the self-interference component given by (3.36) is in this case

$$I = P_{\text{out}} - S = \frac{1}{2}R_d^P(0) = \sum_{\gamma=1}^{+\infty} P_{2\gamma+1}. \tag{3.85}$$

Concerning the PSD, we can write

$$G_y^P(f) = |\alpha_P|^2 G_x^P(f) + G_d^P(f), \tag{3.86}$$

with $G_d^P(f)$ denoting the PSD of the self-interference component, given by

$$G_d^P(f) = \mathcal{F}\{R_d^P(\tau)\} = 2\sum_{\gamma=1}^{+\infty} \frac{P_{2\gamma+1}}{(R_x^P(0))^{2\gamma+1}} f_{2\gamma+1}^G(G_x^P(f)). \tag{3.87}$$

Chapter 4

Accuracy of the Gaussian Approximation for the Evaluation of Nonlinear Effects

As mentioned in Chapter 1, multicarrier signals are very vulnerable to nonlinear distortion effects. As seen in Chapter 3, when the number of subcarriers is high, multicarrier signals have a Gaussian-like nature, which can be used to characterize a multicarrier signal submitted to a nonlinear device, and this characterization can then be employed for performance evaluation of nonlinearly distorted multicarrier signals.

Since practical multicarrier schemes have a finite number of subcarriers, the Gaussian approximation of multicarrier signals might not be accurate, especially when the number of subcarriers is not very high. An accurate characterization of the effects of nonlinear distortion is, therefore, critical to the future of multicarrier techniques.

In this chapter, we present exact characterizations of multicarrier signals submitted to polar memoryless nonlinear devices with characteristics of the form $f(R) = R^{2p+1}$. We will first present exact characterizations for nonlinear functions $f(R) = R^3$ and $f(R) = R^5$, and then present an approximated generalization for higher characteristics. These characterizations will be used to evaluate the accuracy of the Gaussian approximation for a given number of subcarriers.

This chapter is organized as follows: Section 4.1 presents the exact characterization of nonlinearly distorted multicarrier signals and Section 4.2 describes the analytical characterization of nonlinearly distorted multicarrier signals using the Gaussian approximation. A discussion on the accuracy of the Gaussian approximation is made in Section 4.3. Finally, Section 4.4 is concerned with the discussion of results presented in this chapter.

The basic results from this chapter were published in [AD10b, AD12b].

4.1 Exact Characterization of Nonlinearly Distorted Multicarrier Signals

Let us consider the transmission of a nonlinearly distorted multicarrier signal, as depicted in Figure 4.1. Each of the transmitted frequency-domain symbols, $\{\tilde{S}_k; k = 0, 1, \ldots, N - 1\}$, is selected from a given constellation, according to the transmitted data, as seen in Chapter 2. The time-domain block $\{s_n; n = 0, 1, \ldots, N' - 1\} = \text{IDFT}\,\{S_k; k = 0, 1, \ldots, N' - 1\}$, with S_k given by (2.29), can be regarded as a sampled version of the multicarrier burst (2.30) (which includes the oversampling factor $M_{\text{Tx}} = N'/N$) and is given by (2.28)

$$s_n = \frac{1}{\sqrt{N}} \sum_{k=0}^{N-1} S_k e^{j2\pi kn/N'} = \frac{1}{\sqrt{N}} \sum_{k \in \mathcal{K}^{(1)}} S_k e^{j2\pi kn/N'}, \qquad (4.1)$$

where $\mathcal{K}^{(1)} = \{0, 1, \ldots, N - 1\}$.

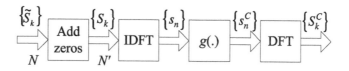

Figure 4.1: Transmitter structure with memoryless nonlinearity device.

We assume the signal is submitted to a polar memoryless nonlinear device (this approach can also be extended to Cartesian memoryless nonlinear devices), as shown in Figure 3.5. In this case, as seen in Subsection 3.1.3, the complex envelope of the signal at the output of the nonlinearity can be written as (3.22). We will assume an odd polar memoryless nonlinear device whose characteristic can be expanded as a power series of the form (3.23). This device operates on the oversampled version of the multicarrier signal, leading to samples

$$s_n^C = \sum_{m=0}^{+\infty} \beta_m s_n^{(2m+1)},$$ (4.2)

where

$$s_n^{(2m+1)} = (s_n)^{m+1}(s_n^*)^m.$$ (4.3)

4.1.1 Nonlinear Characteristic of Order 3

We will first focus on the case of a cubic polar memoryless nonlinear characteristic, i.e., $\beta_m = 0$ for $m \geq 2$. Without loss of generality, we assume $f(R) = R^3$, i.e., $\beta_0 = 0$ and $\beta_1 = 1$ (the extension to other cases is straightforward). Therefore,

$$s_n^C = s_n^{(3)} = s_n s_n^* s_n$$ (4.4)

and these samples can be written as

$$s_n^{(3)} = \frac{1}{\sqrt{N^3}} \sum_{k_1=0}^{N-1} \sum_{k_2=0}^{N-1} \sum_{k_3=0}^{N-1} S_{k_1} S_{k_2}^* S_{k_3} e^{j2\pi(k_1-k_2+k_3)n/N'}$$

$$= \frac{1}{\sqrt{N^3}} \sum_{k^{(3)} \in \mathcal{K}^{(3)}} S_{k_1} S_{k_2}^* S_{k_3} e^{j2\pi(k_1-k_2+k_3)n/N'},$$ (4.5)

with $k^{(3)} = (k_1, k_2, k_3)$ and $\mathcal{K}^{(3)} = \mathcal{K}^{(1)} \times \mathcal{K}^{(1)} \times \mathcal{K}^{(1)}$.

Noting the particular cases

$$k = k_1, k' = k_2 = k_3$$ (4.6a)
$$k = k_3, k' = k_2 = k_1,$$ (4.6b)

we define the sets

$$\mathcal{K}_1^{(3)} = \{(k_1, k_2, k_3) \in \mathcal{K}^{(3)} : k_2 = k_1\}$$ (4.7a)
$$\mathcal{K}_2^{(3)} = \{(k_1, k_2, k_3) \in \mathcal{K}^{(3)} : k_2 = k_3\}$$ (4.7b)

and rewrite samples $s_n^{(3)}$ as a sum of two parcels, i.e.,

$$s_n^{(3)} = \frac{1}{\sqrt{N^3}} \sum_{k^{(3)} \in \mathcal{U}^{(3)}} S_k |S_{k'}|^2 \, e^{j2\pi kn/N'}$$

$$+ \frac{1}{\sqrt{N^3}} \sum_{k^{(3)} \in \mathcal{K}^{(3)} \setminus \mathcal{U}^{(3)}} S_{k_1} S_{k_2}^* S_{k_3} e^{j2\pi(k_1 - k_2 + k_3)n/N'}, \qquad (4.8)$$

with $\mathcal{U}^{(3)} = \mathcal{K}_1^{(3)} \cup \mathcal{K}_2^{(3)}$. Clearly,

$$|\mathcal{K}^{(3)}| = N^3 \qquad (4.9a)$$

$$|\mathcal{K}_1^{(3)}| = |\mathcal{K}_2^{(3)}| = N \qquad (4.9b)$$

and, since

$$|\mathcal{U}^{(3)}| = |\mathcal{K}_1^{(3)} \cup \mathcal{K}_2^{(3)}| = |\mathcal{K}_1^{(3)}| + |\mathcal{K}_2^{(3)}| - |\mathcal{K}_1^{(3)} \cap \mathcal{K}_2^{(3)}| = 2N - 1 \quad (4.10)$$

($|S|$ denotes 'cardinal of set S'), the first summation in (4.8) reduces to

$$\sum_{k^{(3)} \in \mathcal{U}^{(3)}} S_k |S_{k'}|^2 \, e^{j2\pi kn/N'} = (2N - 1) E[|S_k|^2] \sum_{k \in \mathcal{K}^{(1)}} S_k e^{j2\pi kn/N'}.$$

$$(4.11)$$

Hence, the two parcels in equation (4.8) obviously correspond to useful and self-interfering components, i.e.,

$$s_n^{(3)} = \alpha^{(3)} s_n + d_n^{(3)}, \qquad (4.12)$$

with

$$\alpha^{(3)} = \frac{2N - 1}{N} E[|S_k|^2] = \left(2 - \frac{1}{N}\right) E[|S_k|^2] \qquad (4.13)$$

and

$$d_n^{(3)} = \frac{1}{\sqrt{N^3}} \sum_{k^{(3)} \in \mathcal{K}^{(3)} \setminus \mathcal{U}^{(3)}} S_{k_1} S_{k_2}^* S_{k_3} e^{j2\pi(k_1 - k_2 + k_3)n/N'}. \qquad (4.14)$$

Let us define multiplicity of subcarrier k as the number of times we get

$$k_1 - k_2 + k_3 - \dots - k_{2p} + k_{2p+1} = k \qquad (4.15)$$

for all possible $(2p + 1)$-tuples of set $\mathcal{K}^{(2p+1)}$ ($\mathcal{K}^{(n)}$ denotes the n-ary Cartesian product over set $\mathcal{K}^{(1)}$). Letting $M_k^{(2p+1)}$ denote the multiplicity of the kth subcarrier associated to a nonlinear characteristic of type R^{2p+1} and $M_k^{(2p+1 \to 2\gamma+1)}$ denote the multiplicity of that subcarrier associated with the Intermodulation Product (IMP) of order γ, with $\gamma \le p$, produced by the same nonlinear characteristic, it is clear that

$$M_k^{(1)} = \begin{cases} 1, & k = 0, 1, \dots, N - 1 \\ 0, & \text{otherwise} \end{cases} \qquad (4.16)$$

and

$$\sum_{k=0}^{N-1} M_k^{(1)} = N. \tag{4.17}$$

For the signal $s_n^{(3)}$, the multiplicity of subcarrier k is $M_k^{(3)}$, where the block $M^{(3)} = \{M_k^{(3)}; k = -4N+1, -4N+2, \ldots, 5N-2\}$ is the convolution of the augmented blocks $M^{(1)} = \{M_k^{(1)}; k = -N, -N+1, \ldots, 2N-1\}$, i.e.,

$$M^{(3)} = M^{(1)} * M^{(1)} * M^{(1)}. \tag{4.18}$$

The augmented block is obtained by adding $2N$ zeros to the initial block, thus ensuring there is no aliasing when computing the convolution. The block $M^{(3)}$ can be obtained by computing the Inverse Discrete Fourier Transform (IDFT) of the block $\{m_n^{(1)} m_{-n}^{(1)} m_n^{(1)}; n = -N, -N+1, \ldots, 2N-1\}$, with $\{m_n^{(1)}; n = -N, -N+1, \ldots, 2N-1\} = \text{DFT} \{M_k^{(1)}; k = -N, -N+1, \ldots, 2N-1\}$. The generalization of $M^{(2p+1)}$ to any value of p is straightforward.

The discrete convolution $u = x * y$ is given by

$$u_k = \sum_{k'=\text{Max}_1}^{\text{min}_1} x_{k'} y_{k-k'+1}, \tag{4.19}$$

$k = 0, \ldots, 2M-2$, with $\text{Max}_1 = \max(1, k-M+1)$ and $\text{min}_1 = \min(k, M)$, where M is the length of vectors x and y. Consequently, for $v = u * z = x * y * z$, we have

$$
\begin{aligned}
v_k &= \sum_{k''=\text{Max}_2}^{\text{min}_2} u_k z_{k-k''+1} \\
&= \sum_{k''=\text{Max}_2}^{\text{min}_2} \sum_{k'=\text{Max}_1}^{\text{min}_1} x_{k'} y_{k-k'+1} z_{k-k''+1},
\end{aligned} \tag{4.20}
$$

$k = 0, \ldots, 3M-3$, with $\text{Max}_2 = \max(1, k-2M+2)$ and $\text{min}_2 = \min(k, M)$. It follows that

$$M_k^{(2)} = \begin{cases} \dfrac{N}{2} + k, & -\dfrac{N}{2} + 1 \le k \le \dfrac{N}{2} - 1 \\[2mm] \dfrac{3N}{2} - k, & \dfrac{N}{2} \le k \le \dfrac{3N}{2} - 1 \\[2mm] 0, & \text{otherwise} \end{cases} \tag{4.21}$$

and

$$
M_k^{(3)} = \begin{cases}
\dfrac{1}{2}(k+N)(k+N+1), & -N+1 \le k \le -1 \\[2mm]
\dfrac{1}{2}3N(2k+N+1) - (k+N)(k+N+1), & 0 \le k \le N-1 \\[2mm]
\dfrac{1}{2}(2N-k)(2N-k-1), & N \le k \le 2N-2 \\[2mm]
0, & \text{otherwise.}
\end{cases}
$$

$$(4.22)$$

Obviously,

$$
\sum_{k=-N/2+1}^{3N/2-1} M_k^{(2)} = N^2 \tag{4.23}
$$

and

$$
\sum_{k=-N+1}^{2N-2} M_k^{(3)} = N^3. \tag{4.24}
$$

The useful component has multiplicity

$$
M_k^{(3\to1)} = \begin{cases} 2N-1, & 0 \le k \le N-1 \\ 0, & \text{otherwise,} \end{cases} \tag{4.25}
$$

which means that the multiplicity of the self-interference samples is $M_k^{(3\to3)} = M_k^{(3)} - M_k^{(3\to1)}$ and has sum

$$
\sum_{k=-N+1}^{2N-2} M_k^{(3\to3)} = \sum_{k=-N+1}^{2N-2} M_k^{(3)} - \sum_{k=0}^{N-1} M_k^{(3\to1)} = N^3 - 2N^2 + N. \tag{4.26}
$$

The evolution of multiplicities $M_k^{(3)}$, $M_k^{(3\to3)}$ and $M_k^{(3\to1)}$ are depicted in Figure 4.2.

The power of the useful component is

$$
S^{(3)} = P_1^{(3)} = \frac{(\alpha^{(3)})^2}{2} E[|S_k|^2] = \frac{1}{2}\left(2 - \frac{1}{N}\right)^2 E[|S_k|^2]^3 \tag{4.27}
$$

and, recalling that $d_n^{(3)}$ is given by (4.14), the power of the self-interference component is

$$
I^{(3)} = P_3^{(3)} = 2! \frac{N^3 - 2N^2 + N}{N^3} \frac{(E[|S_k|^2])^3}{2}
$$
$$
= \left(1 - \frac{2}{N} + \frac{1}{N^2}\right)(E[|S_k|^2])^3, \tag{4.28}
$$

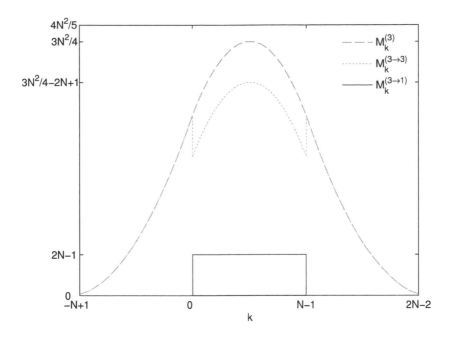

Figure 4.2: Evolution of $M_k^{(3)}$ (dashed line), $M_k^{(3\to3)}$ (dotted line) and $M_k^{(3\to1)}$ (solid line).

where the factor 2! is the number of times we can get repetitions of the values taken by k_1 and k_3.

If $E[|S_k|^2] = 2\sigma^2$, then

$$\alpha^{(3)} = \left(4 - \frac{2}{N}\right)\sigma^2 \qquad (4.29a)$$

$$P_1^{(3)} = \left(4 - \frac{2}{N}\right)^2 \sigma^6 \qquad (4.29b)$$

$$P_3^{(3)} = 8\left(1 - \frac{2}{N} + \frac{1}{N^2}\right)\sigma^6. \qquad (4.29c)$$

For a large number of subcarriers, i.e., $N \gg 1$, these expressions can be approximated by

$$\alpha^{(3)} \approx 4\sigma^2 \tag{4.30a}$$

$$P_1^{(3)} \approx 16\sigma^6 \tag{4.30b}$$

$$P_3^{(3)} \approx 8\sigma^6. \tag{4.30c}$$

Notice these are the same values we get by using the Gaussian approximation results presented in Appendix B, with $f(R) = R^3$.

4.1.2 Nonlinear Characteristic of Order 5

Let us now consider a polar memoryless nonlinear with characteristic of order 5, i.e., $\beta_m = 0$ for $m \geq 3$. Again, without loss of generality, we will assume $f(x) = x^5$, i.e., $\beta_0 = 0$, $\beta_1 = 0$ and $\beta_2 = 1$. Therefore,

$$
\begin{aligned}
s_n^{(5)} &= s_n s_n^* s_n s_n^* s_n \\
&= \frac{1}{\sqrt{N^5}} \sum_{k_1=0}^{N-1} \sum_{k_2=0}^{N-1} \sum_{k_3=0}^{N-1} \sum_{k_4=0}^{N-1} \sum_{k_5=0}^{N-1} \\
&\quad \times S_{k_1} S_{k_2}^* S_{k_3} S_{k_4}^* S_{k_5} e^{j2\pi(k_1 - k_2 + k_3 - k_4 + k_5)n/N'}.
\end{aligned}
\tag{4.31}
$$

In Appendix C, it is shown that we can write $s_n^{(5)}$ as a sum of useful and self-interference components

$$s_n^{(5)} = \alpha^{(5)} s_n + d_n^{(5)}, \tag{4.32}$$

with

$$\alpha^{(5)} = \left(6 - \frac{9}{N} + \frac{4}{N^2}\right) (E[|S_k|^2])^2 \tag{4.33}$$

and self-interference samples given by

$$d_n^{(5)} = d_n^{(5\to3)} + d_n^{(5\to5)}, \tag{4.34}$$

with

$$
d_n^{(5\to3)} = \frac{1}{N} \sum_k M_k^{(5\to3)} E[|S_k|^2]
$$

$$
\cdot \underbrace{\frac{1}{\sqrt{N^3}} \sum_{k'} \sum_{k''} \sum_{k'''} S_{k'} S_{k''}^* S_{k'''} e^{j2\pi(k'-k''+k''')n/N'}}_{\approx d_n^{(3)}}
$$

$$\tag{4.35}$$

and

$$d_n^{(5\to5)} = \frac{1}{\sqrt{N^5}}\sum_{k_1}\sum_{k_2}\sum_{k_3}\sum_{k_4}\sum_{k_5}$$

$$\times S_{k_1} S_{k_2}^* S_{k_3} S_{k_4}^* S_{k_5} e^{j2\pi(k_1-k_2+k_3-k_4+k_5)n/N'}. \tag{4.36}$$

Let $M_k^{(5)}$ be the total multiplicity of subcarrier k, $M_k^{(5\to1)}$ the multiplicity of the useful component, and $M_k^{(5\to3)}$ and $M_k^{(5\to5)}$ the multiplicities of the self-interference components. Using the results from Appendix C, we can write

$$M_k^{(5\to1)} = (6N^2 - 9N + 4)\,M_k^{(1)} \tag{4.37a}$$

$$M_k^{(5\to3)} \approx (6N - 9)M_k^{(3\to3)} + (N - 2)M_k^{(1)} \tag{4.37b}$$

$$M_k^{(5\to5)} \approx M_k^{(5)} - M_k^{(5\to3)} - M_k^{(5\to1)}. \tag{4.37c}$$

The evolution of $M_k^{(5)}$, $M_k^{(5\to1)}$, $M_k^{(5\to3)}$ and $M_k^{(5\to5)}$ is depicted in Figure 4.3.

In case $E[|S_k|^2] = 2\sigma^2$, using the results of Appendix C, we can write the power of the useful and self-interfering components as

$$\alpha^{(5)} = 4\left(6 - \frac{9}{N} + \frac{4}{N^2}\right)\sigma^4 \tag{4.38a}$$

$$P_1^{(5)} = 16\left(6 - \frac{9}{N} + \frac{4}{N^2}\right)^2 \sigma^{10} \tag{4.38b}$$

$$P_3^{(5)} = 32\left(6 - \frac{9}{N}\right)^2\left(1 - \frac{2}{N} + \frac{1}{N^2}\right)\sigma^{10} \tag{4.38c}$$

$$P_5^{(5)} = 192\left(1 - \frac{6}{N} + \frac{15}{N^2} - \frac{17}{N^3} + \frac{7}{N^4}\right)\sigma^{10} \tag{4.38d}$$

and the total power of the self-interference samples is simply $I^{(5)} = P_3^{(5)} + P_5^{(5)}$. When the number of subcarriers is high, the following approximations can be used

$$\alpha^{(5)} \approx 24\sigma^4 \tag{4.39a}$$

$$P_1^{(5)} \approx 576\sigma^{10} \tag{4.39b}$$

$$P_3^{(5)} \approx 1152\sigma^{10} \tag{4.39c}$$

$$P_5^{(5)} \approx 192\sigma^{10} \tag{4.39d}$$

$$I^{(5)} \approx 1344\sigma^{10}. \tag{4.39e}$$

Again, we can notice that these are the same expressions we get by using $f(R) = R^5$ in the results of Appendix B.

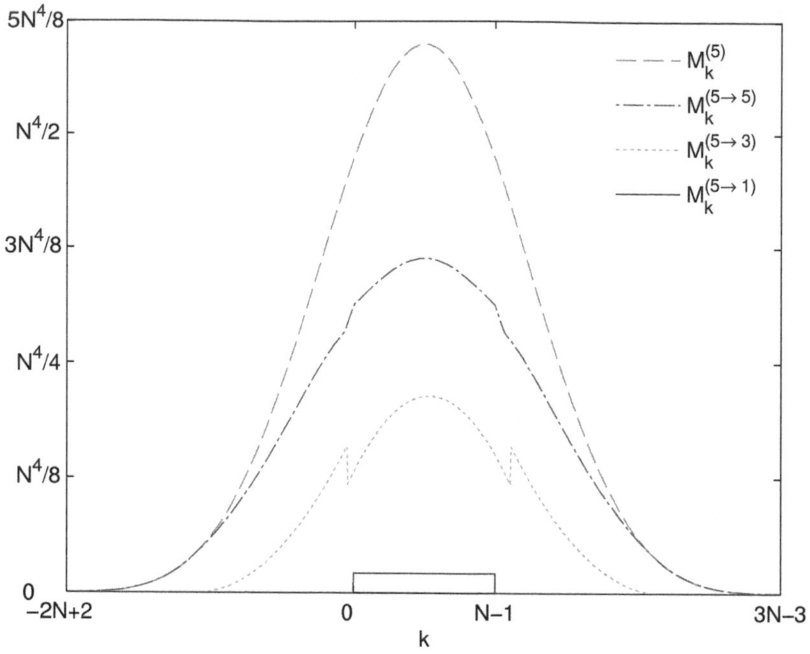

Figure 4.3: Evolution of $M_k^{(5)}$ (dashed line), $M_k^{(5\to5)}$ (dash-dotted line), $M_k^{(5\to3)}$ (dotted line) and $M_k^{(5\to1)}$ (solid line).

4.1.3   Generalization

Using the results in Appendix C, the approach from the previous subsections can be generalized for large values of N. The multiplicities for the signal $s_n^{(2p+1)}$ are approximately given by

$$
M_k^{(2p+1\to2\gamma+1)} \approx
\begin{cases}
\dfrac{p!}{\gamma!(p-\gamma)!}\dfrac{(p+1)!}{(\gamma+1)!}N^{p+\gamma}, & \gamma < p \\[2ex]
M_k^{(2p+1)} - \displaystyle\sum_{l=0}^{p-1} M_k^{(2p+1\to2l+1)}, & \gamma = p,
\end{cases}
\tag{4.40}
$$

and the power of the useful and self-interfering components can be approximated by

$$P_{2\gamma+1}^{(2p+1)} \approx \frac{1}{2} \frac{1}{\gamma!(\gamma+1)!} \left(\frac{p!(p+1)!}{(p-\gamma)!} \right)^2 (E[|S_k|^2])^{2p+1} \qquad (4.41)$$

and, if $p = \gamma$,

$$P_{2p+1}^{(2p+1)} \approx \frac{1}{N^{2p+1}} p!(p+1)! \left(N^{2p+1} \right)^2 \frac{(E[|S_k|^2])^{2p+1}}{2N^{2p+1}}$$

$$\approx \frac{p!(p+1)!}{2} (E[|S_k|^2])^{2p+1}. \qquad (4.42)$$

If $E[|S_k|^2] = 2\sigma^2$, we get

$$P_{2\gamma+1}^{(2p+1)} \approx \frac{1}{\gamma!(\gamma+1)!} \left(\frac{p!(p+1)!}{(p-\gamma)!} \right)^2 2^{2p} \sigma^{4p+2} \qquad (4.43)$$

and

$$P_{2p+1}^{(2p+1)} \approx p!(p+1)! \, 2^{2p} \sigma^{4p+2}, \qquad (4.44)$$

which are the expressions obtained by using the Gaussian approximation (see Appendix B).

4.2 Analytical Characterization of Nonlinearly Distorted Multicarrier Signals Using the Gaussian Approximation

In this section, we will take advantage of the Gaussian nature of multicarrier signals with a large number of subcarriers and the results presented in Chapter 3 for the analytical characterization of nonlinearly distorted transmitted samples. From (3.30), the output of a polar memoryless nonlinear device with a Gaussian input can be written as the sum of two uncorrelated components: a useful one, which is proportional to the input, and a self-interference one, i.e.,

$$s_n^C = \alpha s_n + d_n, \qquad (4.45)$$

where $E[s_n d_{n'}^*] = 0$ and, using (3.79) and (3.80),

$$\alpha = \frac{E[s_n^C s_n^*]}{E[|s_n|^2]} = \frac{E[R f(R)]}{E[R^2]} = \frac{1}{2\sigma^4} \int_0^{+\infty} R^2 f(R) e^{-\frac{R^2}{2\sigma^2}} dR, \qquad (4.46)$$

with $\sigma^2 = E[|s_n|^2]/2$ and $R = |s_n|$. For $f(R) = R^{2p+1}$, the useful component scale factor α is given by

$$\alpha^{(2p+1)} = 2^p (p+1)! \sigma^{2p} \tag{4.47}$$

(see (B.60)).

Using results from Section 3.2, the autocorrelation of the output samples can be expressed as a function of the autocorrelation of the input samples from (3.74), i.e.,

$$R_{s,n-n'}^C = E[s_n^C s_{n'}^{C*}] = 2 \sum_{\gamma=0}^{+\infty} P_{2\gamma+1} \frac{(R_{s,n-n'})^{\gamma+1}(R_{s,n-n'}^*)^{\gamma}}{(R_{s,0})^{2\gamma+1}}, \tag{4.48}$$

with $R_{s,n-n'} = E[s_n s_{n'}^*]$ given by (2.74) and $P_{2\gamma+1}$ denoting the total power associated to the IMP of order $2\gamma+1$, which is given by (3.76). In Appendix B, it is shown that for the particular case of $f(R) = R^{2p+1}$, coefficients $P_{2\gamma+1}$ are given by (B.64) and (B.65)

$$P_{2\gamma+1}^{(2p+1)} = \frac{1}{\gamma!(\gamma+1)!} \left(\frac{p!(p+1)!}{(p-\gamma)!} \right)^2 2^{2p} \sigma^{4p+2} \tag{4.49a}$$

$$P_{2p+1}^{(2p+1)} = p!(p+1)! \, 2^{2p} \sigma^{4p+2}. \tag{4.49b}$$

It can be easily recognized that, as in (3.82), we can write

$$R_{s,n-n'}^C = |\alpha|^2 R_{s,n-n'} + R_{d,n-n'}, \tag{4.50}$$

with

$$R_{d,n-n'} = E[d_n d_{n'}^*] = 2 \sum_{\gamma=1}^{+\infty} P_{2\gamma+1} \frac{(R_{s,n-n'})^{\gamma+1}(R_{s,n-n'}^*)^{\gamma}}{(R_{s,0})^{2\gamma+1}}. \tag{4.51}$$

The average power of the signal at the nonlinearity output is given by (3.84)

$$P_{\text{out}} = \frac{1}{2} E[g^2(R)] = \frac{1}{2\sigma^2} \int_0^{+\infty} R \, g^2(R) \, e^{-\frac{R^2}{2\sigma^2}} \, dR, \tag{4.52}$$

with useful and self-interference components

$$S = P_1 = |\alpha|^2 \sigma^2 \tag{4.53}$$

and

$$I = P_{\text{out}} - S = \frac{1}{2} R_{d,0} = \sum_{\gamma=1}^{+\infty} P_{2\gamma+1}, \tag{4.54}$$

respectively.

Let us now consider the Power Spectral Density (PSD) of the transmitted samples. Having in mind (4.45) and the signal processing chain in Figure 4.1, the frequency-domain samples can obviously be decomposed into useful and nonlinear self-interference components:

$$S_k^C = \alpha S_k + D_k, \tag{4.55}$$

where $\{D_k; k = 0, 1, \ldots, N' - 1\} = \text{DFT}\,\{d_n; n = 0, 1, \ldots, N' - 1\}$. Moreover, $E[d_n] = 0$, leading to $E[D_k] = 0$. To obtain $E[D_k D_{k'}^*]$, we can proceed as follows

$$
\begin{aligned}
E[D_k D_{k'}^*] &= \sum_{n=0}^{N'-1} \sum_{n'=0}^{N'-1} E[d_n d_{n'}^*]\, e^{-j2\pi \frac{kn - k'n'}{N'}} \\
&= \sum_{n=0}^{N'-1} \sum_{n'=0}^{N'-1} R_{d,n-n'}\, e^{-j2\pi \frac{kn - k'n'}{N'}} \\
&= \sum_{n'=0}^{N'-1} \left(\sum_{n=0}^{N'-1} R_{d,n-n'}\, e^{-j2\pi \frac{kn}{N'}} \right) e^{j2\pi \frac{k'n'}{N'}} \\
&= \sum_{n'=0}^{N'-1} G_{D,k}\, e^{-j2\pi \frac{(k-k')n'}{N'}} \\
&= \begin{cases} N' G_{D,k}, & k = k' \\ 0, & \text{otherwise,} \end{cases}
\end{aligned}
\tag{4.56}
$$

where $\{G_{D,k}; k = 0, 1, \ldots, N' - 1\} = \text{DFT}\,\{R_{d,n}; n = 0, 1, \ldots, N' - 1\}$. Similarly, it can be shown that

$$E[S_k^C S_{k'}^{C*}] = \begin{cases} N' G_{S,k}^C, & k = k' \\ 0, & \text{otherwise,} \end{cases} \tag{4.57}$$

with $\{G_{S,k}^C; k = 0, 1, \ldots, N' - 1\} = \text{DFT}\,\{R_{s,n}^C; n = 0, 1, \ldots, N' - 1\}$ and

$$G_{S,k}^C = 2 \sum_{\gamma=0}^{+\infty} \frac{P_{2\gamma+1}}{(R_{s,0})^{2\gamma+1}} \left(G_{S,k}\right)^{2\gamma+1} \tag{4.58}$$

(see (2.63) and (3.77)). Clearly,

$$G_{S,k}^C = |\alpha|^2 G_{S,k} + G_{D,k}, \tag{4.59}$$

with

$$G_{D,k} = 2 \sum_{\gamma=1}^{+\infty} \frac{P_{2\gamma+1}}{(R_{s,0})^{2\gamma+1}} \left(G_{S,k}\right)^{2\gamma+1}. \tag{4.60}$$

Table 4.1: Comparison of $\alpha^{2p+1}/\sigma^{2p}$ for different values of N.

N	$\alpha^{(3)}$			$\alpha^{(5)}$			
	approx	exact	simul	approx1	approx2	exact	simul
16	4	3.875	3.873	24	21.750	21.813	21.869
64	4	3.969	3.972	24	23.438	23.441	23.457
256	4	3.992	3.991	24	23.859	23.860	23.872

Moreover, it is usual to consider D_k exhibits quasi-Gaussian characteristics for any k provided that the number of subcarriers is high enough. This is true if the number of contributions on each subcarrier is much greater than one, due to the central limit theorem.

4.3 Accuracy of the Gaussian Approximation

From the comparison of exact results and the ones obtained using the Gaussian approximation, it is clear that this approximation leads to errors. In this section, we present some results concerning these two approaches.

The Gaussian approximation leads to an error of $2/N$ and $36/N - 16/N^2$ on the evaluation of $\alpha^{(3)}$ and $\alpha^{(5)}$ values, respectively. This is about 3.2% for $N = 16$, 0.8% for $N = 64$, and 0.2% for $N = 256$ in the case of $\alpha^{(3)}$ and 10% for $N = 16$, 2.4% for $N = 64$ and 0.6% for $N = 256$ in the case of $\alpha^{(5)}$. Table 4.1 shows values obtained using the exact characterization and the Gaussian approximation for $\alpha^{(3)}$ and $\alpha^{(5)}$ obtained using the exact expressions (4.29) and (4.38), and the approximations (4.30) and (4.39) for $N = 16$, $N = 64$ and $N = 256$. In the case of $\alpha^{(5)}$, two approximations are used, the one obtained from (4.38) just by neglecting the $16/N^2$ term and (4.39). Values obtained by simulation for $\alpha^{(3)}$ and $\alpha^{(5)}$ are very close to expressions (4.29) and (4.38). A comparison of the exact characterization and Gaussian approximation values for $\alpha^{(3)}$ and $\alpha^{(5)}$ obtained using the exact expressions (4.29) and (4.38), and the approximations (4.30) and (4.39), respectively, as a function of the number of subcarriers, is shown in Figure 4.4. It can be seen that for values of N greater then 100, there are no significant differences between the exact and the approximated values.

From the comparison of equations (4.29) and (4.30), it is clear that with $f(R) = R^3$, the Gaussian approximation leads to an error of about

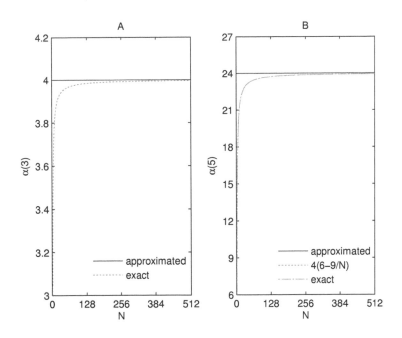

Figure 4.4: Evolution of $\alpha^{(3)}$ (A) and $\alpha^{(5)}$ (B), using exact and approximated expressions.

$(1 - 1/(2N))^2/(1 - 2/N + 1/N^2)$ on the evaluation of the Signal-to-Interference Ratio (SIR) levels, with

$$\text{SIR}_k = \frac{\alpha^2 E[|S_k|^2]}{E[|D_k|^2]}. \tag{4.61}$$

If $N^2 \gg 1$ (say $N > 10$), the terms $1/N^2$ can be neglected and the error is approximately $(1 - 1/(2N))^2/(1 - 2/N)$, which is about 7.3% for $N = 16$, 1.6% for $N = 64$, and 0.4% for $N = 256$. In the case of $f(R) = R^5$, looking at equations (4.38) and (4.39) and neglecting the terms $1/N^m$, with $m \geq 2$, we find the error on SIR levels is approximately $(6 - 9/N)^2/$ $[(1 - 3/(2N))^2(1 - 2/N) + (1 - 6/N)]$. This is about 16.5% for $N = 16$, 3.5% for $N = 64$, and 0.9% for $N = 256$. Figures 4.5 and 4.6 illustrate this by showing the evolution of SIR levels for the Gaussian approximation and the exact characterizations and a different number of subcarriers. The evolution of $E[\text{SIR}_k]$, for Gaussian approximation and exact characterization as a function of the number of subcarriers for $f(R) = R^3$ and $f(R) = R^5$, is depicted in Figure 4.7. As for $\alpha^{(3)}$ and $\alpha^{(5)}$, it can

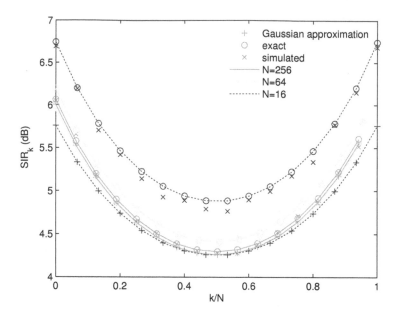

Figure 4.5: Evolution of SIR$_k$, with $f(R) = R^3$ for the Gaussian approximation and the exact characterization with different number of subcarriers.

be seen that if $N \geq 100$, there are no significant differences between the expected values of the SIR levels for the two characterizations.

It is usual to consider D_k exhibits quasi-Gaussian characteristics for any k, provided that the number of subcarriers is high enough. This is true if the number of contributions on each subcarrier is much greater than one, due to the central limit theorem. For $f(R) = R^3$, the number of contributions on subcarrier k is $M_k^{(3 \to 3)} = M_k^{(3)} - M_k^{(3 \to 1)}$, which has approximately N^2 terms, so samples D_k are Gaussian if $N^2 \gg 1$ (say $N > 10$). However, the number of contributions grows with k^2 (see (4.22)), being very low at the edges of the band ($k \approx -N$ or $k \approx 2N$). This implies the Gaussian approximation is not accurate at the edges of the band. The same happens with $f(R) = R^5$, since the number of contributions on subcarrier k is $M_k^{(5)} - M_k^{(5 \to 1)}$ (see (4.37)), which depends on $M_k^{(3 \to 3)}$. A comparison of $E[|D_k|^2]$ values obtained using the Gaussian approximation, exact characterization and simulations is shown in Figures 4.8 and 4.9.

As an example, we present some results for the Solid State Power Am-

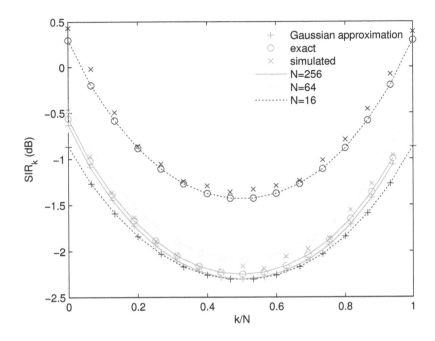

Figure 4.6: Evolution of SIR$_k$, with $f(R) = R^5$ for the Gaussian approximation and the exact characterization with different number of subcarriers.

plifier (SSPA), which includes clipping as a particular case. Its AM/AM and AM/PM conversion functions are, respectively, given by (3.26) and (3.27). The series expansion for an SSPA with $s_M = 2$ and $q = 1$ is

$$A(R) = A_M \left(\frac{1}{2}R - \frac{1}{16}R^3 + \frac{3}{256}R^5 - \frac{5}{2048}R^7 + \frac{35}{65536}R^9 \right) \quad (4.62)$$

and with $s_M = 2$ and $q = 2$ is

$$A(R) = A_M \left(\frac{1}{2}R - \frac{1}{128}R^5 + \frac{5}{16384}R^9 \right). \quad (4.63)$$

As expected, we find that the dominant terms are R and R^3, term R^5 has little influence on the approximation, and the remaining terms have almost no significance. Similar results can be obtained for other values of s_M and q. Figures 4.10 and 4.11 show SIR levels and $E[|D_k|^2]$ values obtained using the Gaussian approximation and simulations for an SSPA with $q = 2$ and $s_M/\sigma = 2.0$.

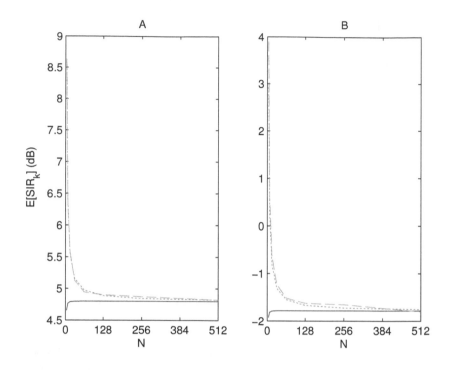

Figure 4.7: Impact of the number of subcarriers on $E[\mathrm{SIR}_k]$, with $f(R) = R^3$ (A) and $f(R) = R^5$ (B) for the Gaussian approximation (solid line), the exact characterization (dotted line) and simulated results (dashed line).

4.4　Discussion

In this chapter, we studied the accuracy of using the Gaussian approach to evaluate the impact of polar memoryless nonlinear devices on multicarrier signals. It was shown that even for a reduced number of subcarriers, a multicarrier signal submitted to a nonlinear device can be decomposed in useful and self-interference components. In the cases of nonlinear characteristics of order 3 and 5, exact expressions for the power of these components were derived and compared with the ones obtained using the Gaussian approximation approach.

From the presented analysis, we concluded that the Gaussian approximation is very good for large values of N, as expected. Values found for the useful component scale factor α are also very accurate. However, the PSD of the self-interference component is overestimated, especially

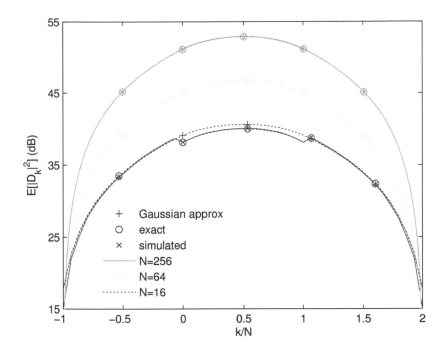

Figure 4.8: Comparison of Gaussian approximation, exact and simulated values of $E[|D_k|^2]$, with $f(R) = R^3$.

in the in-band region, leading to slightly pessimistic SIR levels, with an error inversely proportional to N. With respect to the Gaussian approximation of the self-interference component at the subcarrier level, it is very accurate whenever N^2 is high, except at the edge of the band.

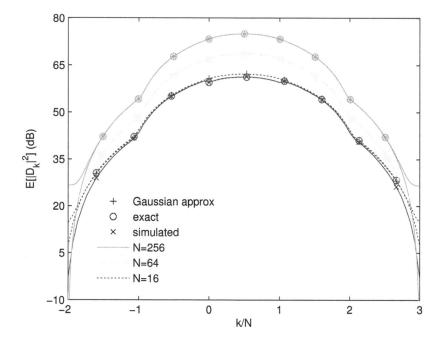

Figure 4.9: Comparison of Gaussian approximation, exact and simulated values of $E[|D_k|^2]$, with $f(R) = R^5$.

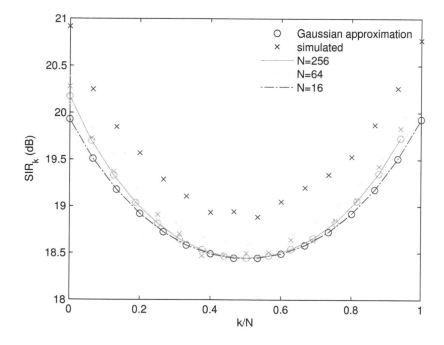

Figure 4.10: Comparison of Gaussian approximation and simulated values of SIR_k, for an SSPA with $q = 2$ and $s_M/\sigma = 2.0$.

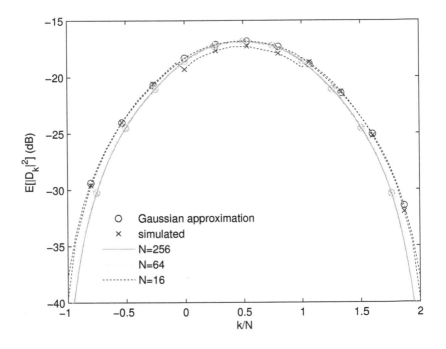

Figure 4.11: Comparison of Gaussian approximation and simulated values of $E[|D_k|^2]$ for an SSPA with $q = 2$ and $s_M/\sigma = 2.0$.

Chapter 5

Nonlinear Effects in Specific Multicarrier Systems

This chapter is dedicated to the application of some of the results presented before to several scenarios where nonlinear distortion in multicarrier systems occurs. For this purpose, we include appropriate statistical characterizations of the signal along the transmission chain on each scenario. These characterizations take advantage of the Gaussian behaviour of the complex envelope of multicarrier signals with a high number of subcarriers, and employ the results on Gaussian signals and memoryless nonlinearities presented in Chapter 3. These statistical characterizations can then be used for performance evaluation of given nonlinear characteristics in a simple and computationally efficient way, as well as to its optimization.

Multicarrier signals have high envelope fluctuations, making them prone to nonlinear distortion effects. In Section 5.1, we describe general clipping and filtering techniques that deal with this problem, thus allowing a more efficient power amplification processing.

One of the problems associated with multicarrier signals is the numerical accuracy required by the Discrete Fourier Transform (DFT)/Inverse Discrete Fourier Transform (IDFT) operations, which can have a significant impact on the transmission performance, especially when large constellations are employed. This accuracy can be modeled as appropriate quantization effects associated to the input or the output of each DFT/IDFT computation. In Section 5.2, we study the impact of quantization effects on the complex envelope of multicarrier signals. These quantization effects occur at both the transmitter and the receiver and are associated with the numerical accuracy of the DFT computations.

Software radio concept is a topic of widespread interest in wireless cellular systems. The Analog-to-Digital Converter (ADC) of a software radio base station should be able to sample and quantize the wideband signal associated with the combination of a large number of users, possibly with different bands and different powers. Therefore, we will need a high sampling ratio, which is inherent to high bandwidths, and a quantizer with a large Signal-to-Noise Ratio (SNR). Due to these requirements, the ADC is as key component of any software radio architecture [Wal99]. As seen in Section 2.3, software radio can be viewed as a multicarrier scheme. Thus, the analytical techniques presented in Chapter 3 can be used for the evaluation of the ADC quantization requirements. Section 5.3 presents an analytical tool for evaluating the quantization effects in software radio architectures. For this purpose, we take advantage of the Gaussian behaviour of the multi-band, multi-user signals at the input of the wideband ADC. As in Section 5.2, the quantizer is modeled as a Cartesian memoryless nonlinear device. The presented approach is therefore similar to the one present in the previous section. This characterization is then used for the performance evaluation and optimization of quantization characteristics of the ADC.

There are several subcarrier parameters available for adjustment on multicarrier systems, and one of the most popular choices is subcarrier signal constellation size. The process of adjusting this parameter, called adaptive bit loading, involves an algorithm that adjusts the number of bits per subcarrier (and corresponding constellation size) according to the channel conditions, i.e., the transmitted number of bits is not equal across all subcarriers. Another parameter that can be adjusted is the energy assigned to each subcarrier. It can be chosen according to the number of bits and channel attenuation on a particular subcarrier. Consequently, adaptive bit loading has the potential to achieve data transmissions that are very spectrally efficient [HH87, LC97, KRJ00, LSC07]. In Section 5.4, we study the impact of nonlinear distortion effects on adaptive multicarrier systems. An analytical statistical characterization of the transmitted signals is included and used for an analytical performance evaluation in the presence of strong nonlinear distortion effects and to define enhanced loading algorithms taking into account nonlinear distortion issues.

Current and evolving standards for broadband wireless systems have adopted or are considering Orthogonal Frequency Division Multiple Access (OFDMA) as the multiple access technology for the air interface. The impact of nonlinear distortion effects in the performance of these systems is therefore a relevant topic. Section 5.5 deals with the analytical evaluation of nonlinear distortion effects on OFDMA signals. We consider nonlinear distortion effects that are inherent to nonlinear signal processing techniques for reducing the Peak-to-Mean Envelope Power Ratio (PMEPR) of the transmitted signals (as the ones proposed in [DG04] for conventional Orthogonal Frequency Division Multiplexing (OFDM) schemes). For this purpose, we take advantage of the Gaussian-like nature of OFDMA signals to extend the results of [DG04, BC00, DTV00] to OFDMA schemes. Our results allow an analytical spectral characterization of the transmitted signals, as well as the computation of the nonlinear interference levels on the received signals. They can also be used to compute the corresponding Bit Error Rate (BER). This allows an efficient approach for studying aspects such as the type of nonlinear device, the impact of the system load (fraction of subcarriers used), the carrier assignment schemes (continuous, randomly spaced or regularly spaced subcarriers), etc.

Each section is followed by a set of performance results. Finally, Section 5.6 is concerned with the conclusions and final remarks of this chapter.

5.1 Clipping and Filtering Techniques

There are many methods proposed to reduce the envelope fluctuations of OFDM signals which, as a consequence, allow a more efficient power amplification processing. These include coding schemes to avoid high amplitude peaks [JW96] and Partial Transmit Sequences (PTS) techniques [MBFH97, MH97, CS99]. However, the simplest and most flexible techniques for reducing the envelope fluctuations of OFDM signals involve a nonlinear clipping operation [OL95, LC98, MR98, DG97, DG00, DG01]. In this section, we describe general clipping and filtering techniques.

5.1.1 Transmitter Structure

The basic transmitter structure is depicted in Figure 5.1. The block of time-domain samples associated with a given OFDM block is generated, as seen in Section 2.2. An augmented frequency-domain block $\{S_k; k = 0, 1, \ldots, N' - 1\}$, with $N' = NM_{\text{Tx}}$ for a selected $M_{\text{Tx}} \geq 1$, is formed by adding $N' - N$ zeros to the original frequency-domain block $\{\tilde{S}_k; k = 0, 1, \ldots, N - 1\}$, directly related to data (see (2.29)). The IDFT of this frequency-domain block is computed, leading to the block $\{s_n; n = 0, 1, \ldots, N' - 1\}$. Each time-domain sample, s_n, is submitted to a clipping operation, leading to a modified sample s_n^C. We consider two types of clipping:

◼ Cartesian clipping, which separately operates on the real and the imaginary parts of each complex sample s_n, i.e.,

$$s_n^C = g_{\text{clip}}(\text{Re}\{s_n\}) + jg_{\text{clip}}(\text{Im}\{s_n\}), \tag{5.1}$$

where $g_{\text{clip}}(x)$ is given by (3.3)

$$g_{\text{clip}}(x) = \begin{cases} -s_M, & x < -s_M \\ x, & |x| \leq s_M \\ s_M, & x > s_M, \end{cases} \tag{5.2}$$

with a the clipping level $s_M = \frac{x_M}{\sqrt{2}}$, where x_M denotes the maximum envelope at the clipping output (as described in Section 3.1).

◼ Envelope clipping, which operates on each complex sample s_n, i.e.,

$$s_n^C = f(|s_n|) \, e^{j \arg(s_n)}, \tag{5.3}$$

with $f(R)$ given by (3.25)

$$f(R) = \begin{cases} R, & R \leq s_M \\ s_M, & x > s_M, \end{cases} \tag{5.4}$$

with s_M denoting the clipping level (see Subsection 3.1.3).

A DFT operation brings the nonlinearly modified block back to the frequency domain, where a linear shaping operation is performed by a multiplier bank with selected coefficients $F_k, k = 0, 1, \ldots, N' - 1$. A frequency-domain-filtered block $\{S_k^{CF} = F_k S_k^C; k = 0, 1, \ldots, N' - 1\}$ is obtained, with

$$S_k^C = \sum_{n=0}^{N'-1} s_n^C \, e^{-j2\pi \frac{kn}{N'}}. \tag{5.5}$$

Since $S_k^{CF} = 0$ for a range of values of k (namely due to $F_k = 0$), a reduced block $\{S_k^{\text{Tx}}; k = 0, 1, \ldots, N'' - 1\}$, with $N \leq N'' \leq N'$, can be derived from $\{S_k^{CF} = F_k S_k^C; k = 0, 1, \ldots, N' - 1\}$, with

$$S_k^{\text{Tx}} = \begin{cases} S_k^{CF}, & 0 \leq k \leq N''/2 - 1 \\ S_{k+N'-N''}^{CF}, & N''/2 \leq k \leq N'' - 1 \end{cases} \tag{5.6}$$

(it is assumed that N'' is a power of two). A second IDFT converts this modified frequency-domain block into the time domain, leading to the block $\{s_n^{\text{Tx}}; n = 0, 1, \ldots, N'' - 1\}$.

The multicarrier burst to be transmitted is then obtained, within the transmitter structure of Figure 5.1, by appending the required cyclic extension to $\{s_n^{\text{Tx}}; n = 0, 1, \ldots, N'' - 1\}$, possibly modifying the amplitude of the samples at the burst tails according to a windowing procedure, and generating the analog real and imaginary signal components through digital to analog conversion and selected low-pass reconstruction filtering.

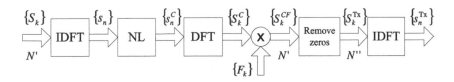

Figure 5.1: OFDM transmission chain with clipping blocks.

Appending $N' - N$ zeros to each initial frequency-domain block prior to computing the required IDFT is a well-known OFDM implementation technique, which is equivalent to oversampling, by a factor $M_{\text{Tx}} = N'/N$, the 'ideal' OFDM burst. The subsequent nonlinear operation proposed here is crucial for reducing the envelope fluctuations, whereas the frequency-domain filtering using the set $\{F_k; k = 0, 1, \ldots, N' - 1\}$ can provide a complementary filtering effect (of course, with some regrowth of the envelope fluctuations). The reduced size that is allowed for the last IDFT essentially means a reduced computational effort; the removal of $N' - N''$ subcarriers with zero amplitude can be regarded as corresponding to a decimation in the time domain.

For a given input block size N, a careful selection of M_{Tx}, the nonlinear characteristic (for a given input level) and $\{F_k; k = 0, 1, \ldots, N' - 1\}$ ensures reduced envelope fluctuations while maintaining low out-of-band radiation levels.

The selection of either a Cartesian or a polar nonlinearity, for PMEPR reduction purposes, defines the choice of a subclass of signal processing schemes within the wide class signal processing considered here.

It should be mentioned that this wide class of signal processing schemes (following and extending our original proposal in [DG00]) includes, as specific cases, signal processing schemes proposed by other authors in the meantime: this is the case of [OI00] ($M_{\text{Tx}} = 1$ and a polar nonlinearity that operates as an envelope clipper). This is also the case of [Arm01] and [OI01], where the same envelope clipping is adopted when assuming $M_{\text{Tx}} \geq 1$, with $F_k = 1$ for the N in-band subcarriers and $F_k = 0$ for the $N' - N$ out-of-band ones.

It should also be mentioned that a more sophisticated technique, allowing improved PMEPR-reducing results, could be simply developed on the basis of the signal processing approach analyzed here. Such technique consists of repeatedly using, in an iterative way, the signal processing chain that leads from $\{S_k\}$ to $\{S_k^{CF}\}$ in Figure 5.1 (of course, $\{S_k^{(l)}\} = \{S_k^{CF(l-1)}\}$ for $l > 1$ and $\{S_k^{(1)}\} = \{S_k\}$, where each upperscript concerns a given iteration). The iterative techniques of [Arm02, DG03b] correspond to the particular case where the nonlinear operation is an envelope clipping and the frequency-domain filtering is characterized by $F_k = 0$ for the $N' - N$ out-of-band subcarriers, with $F_k = 1$ in-band.

5.1.2 Signal Characterization along the Signal Processing Chain

This section shows how to obtain an appropriate statistical characterization for the modified block of frequency-domain samples $\{S_k^{\text{Tx}}; k = 0, 1, \ldots, N'' - 1\}$ that replaces the block $\{S_k; k = 0, 1, \ldots, N' - 1\}$ of conventional OFDM schemes. For this purpose, we need to characterize statistically the blocks along the signal processing chain.

Time-Domain Block at the Input to the Nonlinearity

It is assumed that $E[S_k] = 0$ and $E[S_k S_{k'}^*] = 2\sigma_S^2 \delta_{k,k'}$ ($\delta_{k,k'} = 1$ for $k = k'$ and 0 otherwise), with $\sigma_S^2 = \frac{1}{2} E\left[|S_k|^2\right]$ ($E[\cdot]$ denotes 'ensemble average'), although this analytical approach can easily be extended to other cases, such as the ones where different powers are assigned to different subcarriers [AD10a, AD12a]. Therefore, it can be easily demonstrated that $E[s_n] = 0$ and

$$E[s_n s_{n'}^*] = R_{s,n-n'} = \frac{1}{(N')^2} \sum_{k=0}^{N'-1} G_{S,k} \, e^{j2\pi \frac{k(n-n')}{N'}}, \qquad (5.7)$$

$n, n = 0, 1, \ldots, N' - 1$, i.e., with $\{R_{s,n}; n = 0, 1, \ldots, N' - 1\} = \frac{1}{N'}$ IDFT $\{G_{S,k}; k = 0, 1, \ldots, N' - 1\}$. The variance of both $\text{Re}\{s_n\}$ and $\text{Im}\{s_n\}$ is

$$\sigma^2 = \frac{1}{2} E[|s_n|^2] = \frac{1}{2} R_{s,0} = \frac{1}{2(N')^2} \sum_{k=0}^{N'-1} G_{S,k}. \qquad (5.8)$$

When the number of subcarriers is high ($N \gg 1$) the time-domain coefficients s_n can be approximately regarded as samples of a zero-mean complex Gaussian process [DW01] with autocorrelation given by (5.7).

Time-Domain Block at the Output of the Nonlinearity

In the following, we take advantage of the quasi-Gaussian nature of the samples s_n for obtaining the statistical characterization of the time-domain block at the output of the nonlinearity. It was seen in Chapter 3 that the output of a memoryless nonlinear device with a Gaussian input can be written as the sum of two uncorrelated components: an useful one, proportional to the input, and a self-interference one. This happens for Cartesian and polar nonlinearities. Hence, for both classes of nonlinearities, we can write

$$s_n^C = \alpha^C s_n + d_n^C, \tag{5.9}$$

where $E[s_n d_{n'}^{C*}] = 0$. This means that the nonlinearly modified samples can be decomposed into uncorrelated useful and self-interference components. The scaling factor α^C is given by (3.31) for Cartesian nonlinearities

$$\alpha^C = \frac{E[x g_{\text{clip}}(x)]}{E[x^2]} = \frac{1}{\sqrt{2\pi}\sigma^3} \int_{-\infty}^{+\infty} x \, g_{\text{clip}}(x) \, e^{-\frac{x^2}{2\sigma^2}} \, dx, \tag{5.10}$$

and by (3.80) for polar nonlinearities

$$\alpha^C = \frac{E[R f(R)]}{E[R^2]} = \frac{1}{2\sigma^4} \int_0^{+\infty} R^2 f(R) \, e^{-\frac{R^2}{2\sigma^2}} \, dR. \tag{5.11}$$

The average power of the useful component is $S^C = |\alpha^C|^2 \sigma^2$, and the average power of the self-interference component is given by $I^C = P_{\text{out}}^C - S^C$, where P_{out}^C denotes the average power of the signal at the nonlinearity output. P_{out}^C is given by (3.34)

$$P_{\text{out}}^C = E[g_{\text{clip}}^2(x)] = \frac{1}{\sqrt{2\pi}\sigma} \int_{-\infty}^{+\infty} g_{\text{clip}}^2(x) \, e^{-\frac{x^2}{2\sigma^2}} \, dx \tag{5.12}$$

for Cartesian nonlinearities, and by (3.84)

$$P_{\text{out}}^C = \frac{1}{2\sigma^2} E[f^2(R)] = \frac{1}{2} \int_0^{+\infty} R f^2(R) \, e^{-\frac{R^2}{2\sigma^2}} \, dR \tag{5.13}$$

for polar nonlinearities.

For a Cartesian nonlinearity, it was shown in Section 3.3 that the autocorrelation of the output samples can be expressed as a function of the autocorrelation of the input samples in the following way:

$$
R^C_{s,n-n'} = E[s^C_n s^{C*}_{n'}]
$$

$$
= 2 \sum_{\gamma=0}^{+\infty} P^C_{2\gamma+1} \frac{(\mathrm{Re}\{R_{s,n-n'}\})^{2\gamma+1} + j(\mathrm{Im}\{R_{s,n-n'}\})^{2\gamma+1}}{(R_{s,0})^{2\gamma+1}},
$$

$$(5.14)$$

where $R_{s,n-n'}$ is given by (5.7), and the coefficient $P^C_{2\gamma+1}$ denotes the total power associated to the Intermodulation Product (IMP) of order $2\gamma + 1$, given by (3.51).

In case of a polar nonlinearity, it was shown in Section 3.4 that the autocorrelation of the output samples can be expressed as a function of the autocorrelation of the input samples from (3.74)

$$
R^C_{s,n-n'} = E[s^C_n s^{C*}_{n'}] = 2 \sum_{\gamma=0}^{+\infty} P^C_{2\gamma+1} \frac{(R_{s,n-n'})^{\gamma+1}(R^*_{s,n-n'})^{\gamma}}{(R_{s,0})^{2\gamma+1}}, \quad (5.15)
$$

where, once again, the coefficient $P^C_{2\gamma+1}$ denotes the total power associated with the IMP of order $2\gamma + 1$ and is given by (3.76).

Since

$$
R^C_{s,n-n'} = |\alpha^C|^2 R_{s,n-n'} + E[d^C_n d^{C*}_{n'}], \qquad (5.16)
$$

it can be easily recognized that $P_1 = |\alpha^C|^2 \sigma^2$ and

$$
R^C_{d,n-n'} = E[d^C_n d^{C*}_{n'}]
$$

$$
= 2 \sum_{\gamma=1}^{+\infty} P^C_{2\gamma+1} \frac{(\mathrm{Re}\{R_{s,n-n'}\})^{2\gamma+1} + j\,(\mathrm{Im}\{R_{s,n-n'}\})^{2\gamma+1}}{(R_{s,0})^{2\gamma+1}}
$$

$$(5.17)$$

for Cartesian nonlinearities, and

$$
R^C_{d,n-n'} = E[d^C_n d^{C*}_{n'}] = 2 \sum_{\gamma=1}^{+\infty} P^C_{2\gamma+1} \frac{(R_{s,n-n'})^{\gamma+1}(R^*_{s,n-n'})^{\gamma}}{(R_{s,0})^{2\gamma+1}} \quad (5.18)
$$

for polar nonlinearities. The total power of the self-interference term is

$$
I^C = \sum_{\gamma=1}^{+\infty} P^C_{2\gamma+1} = P^C_{\text{out}} - S^C. \qquad (5.19)
$$

This method for statistical characterization of the transmitted blocks is quite appropriate whenever the power series in (5.17) and (5.18) can be reasonably truncated while ensuring an accurate computation. However, for strongly nonlinear conditions, the required number of terms becomes very high. In such cases, one can simplify the computation as explained below.

When $\gamma \gg M_{\text{Tx}}$,

$$\left(\text{Im}\left\{\frac{R_{s,n-n'}}{R_{s,0}}\right\}\right)^{2\gamma+1} \approx 0 \qquad (5.20)$$

and

$$\left(\frac{R_{s,n-n'}}{R_{s,0}}\right)^{2\gamma+1} \approx \left(\text{Re}\left\{\frac{R_{s,n-n'}}{R_{s,0}}\right\}\right)^{2\gamma+1} \approx \delta_{n,n'}, \qquad (5.21)$$

which means that the frequency-domain distribution of the power associated to a given IMP, $\{G_{S,k}^{2\gamma+1}; \ k = 0, 1, \ldots, N' - 1\} = \text{DFT} \{(R_{s,n}/R_{s,0})^{2\gamma+1}; n = 0, 1, \ldots, N' - 1\}$, is almost constant when $\gamma \gg M_{\text{Tx}}$, leading to

$$\frac{G_{S,k}^{2\gamma+1}}{G_{S,0}^{2\gamma+1}} \approx 1, \qquad (5.22)$$

for $\gamma \gg M_{\text{Tx}}$ (see Figure 5.2).

As a consequence, the contribution of the $(2\gamma + 1)$th IMP to the output auto-correlation can be approximated by $2P_{2\gamma+1}^{C}\delta_{n,n'}$, leading to

$$R_{s,n-n'}^{C} \approx 2\sum_{\gamma=0}^{\gamma_{\max}} P_{2\gamma+1}^{C} \frac{(\text{Re}\{R_{s,n-n'}\})^{2\gamma+1} + j(\text{Im}\{R_{s,n-n'}\})^{2\gamma+1}}{(R_{s,0})^{2\gamma+1}}$$

$$+ 2P_{2\gamma_{\max}+1}^{C,+\infty}\delta_{n,n'} \qquad (5.23)$$

$(\gamma_{\max} \gg M_{\text{Tx}})$ for a Cartesian nonlinearity, with

$$P_{2\gamma_{\max}+1}^{C,+\infty} = \sum_{\gamma=\gamma_{\max}+1}^{+\infty} P_{2\gamma+1}^{C} = I^{C} - \sum_{\gamma=1}^{\gamma_{\max}} P_{2\gamma+1}^{C}, \qquad (5.24)$$

where I^{C} denotes the average power of the self-interference component. For polar nonlinearities,

$$R_{s,n-n'}^{C} \approx 2\sum_{\gamma=0}^{\gamma_{\max}} P_{2\gamma+1}^{C} \frac{(R_{s,n-n'})^{\gamma+1}(R_{s,n-n'}^{*})^{\gamma}}{(R_{s,0})^{2\gamma+1}} + 2P_{2\gamma_{\max}+1}^{C,+\infty}\delta_{n,n'}, \qquad (5.25)$$

with $P_{2\gamma_{max}+1}^{C,+\infty}$ also given by (5.24).

This means that, besides the computation of $I^{C} = P_{\text{out}}^{C} - S^{C}$, we just have to calculate the terms corresponding to the first γ_{max} IMPs.

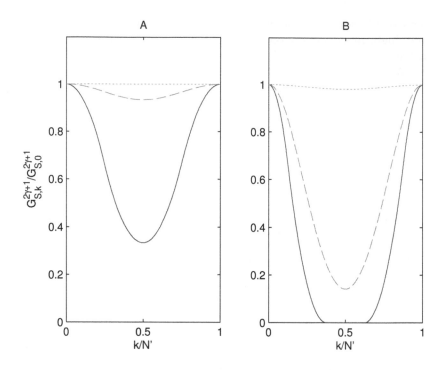

Figure 5.2: Evolution of $|G_{s,2\gamma+1}(k)/G_{s,2\gamma+1}(0)|$ for $M_{\mathbf{Tx}} = 2$ (A) and $M_{\mathbf{Tx}} = 4$ (B) with $\gamma = 1$ (solid line), $\gamma = 4$ (dashed line) and $\gamma = 25$ (dotted line).

Final Frequency-Domain and Time-Domain Blocks

Having in mind (5.9) and the signal processing chain in Figure 5.1, the frequency-domain block $\{S_k^{CF} = S_k^C F_k; k = 0, 1, \ldots, N' - 1\}$ can obviously be decomposed into useful and self-interference components:

$$S_k^{CF} = \alpha^C S_k F_k + D_k^C F_k, \qquad (5.26)$$

where $\{D_k^C; k = 0, 1, \ldots, N' - 1\}$ denotes the DFT of $\{d_n^C; n = 0, 1, \ldots, N' - 1\}$.

Clearly, $E[D_k^C] = 0$ and

$$E[D_k^C D_{k'}^{C*}] = \sum_{n=0}^{N'-1} \sum_{n'=0}^{N'-1} E[d_n^C d_{n'}^{C*}] e^{-j2\pi \frac{kn - k'n'}{N'}} = N' G_{D,k}^C \delta_{k,k'}, \quad (5.27)$$

$k, k' = 0, 1, \ldots, N' - 1$, where $\{G_{D,k}^C; k = 0, 1, \ldots, N' - 1\}$ denotes the

DFT of the block $\{R^C_{d,n}; n = 0, 1, \ldots, N' - 1\}$. Similarly, it can be shown that

$$E[S^C_k S^{C*}_{k'}] = N' G^C_{S,k} \delta_{k,k'}, \qquad (5.28)$$

where $\{G^C_{S,k} = |\alpha|^2 G_{S,k} + G_{D,k}; k = 0, 1, \ldots, N' - 1\}$ denotes the DFT of $\{R^C_{s,n}; k = 0, 1, \ldots, N' - 1\}$ (given by (5.14) or (5.15), according to the type of nonlinearity), with $\{G_{S,k}; k = 0, 1, \ldots, N' - 1\} = \text{DFT}$ $\{R_{s,n}; n = 0, 1, \ldots, N' - 1\}$. Therefore, $E[S^{CF}_k S^{CF*}_{k'}] = 0$ for $k \neq k'$, and $E[|S^{CF}_k|^2] = |F_k|^2 E[|S^C_k|^2] = N'|F_k|^2 G^C_{S,k}$.

By employing the statistical characterization of the frequency-domain block to be transmitted, one can calculate the Signal-to-Interference Ratio (SIR) for each subcarrier, which is given by

$$\text{SIR}_k = \frac{E[|\alpha^C S_k|^2]}{E[|D^C_k|^2]}. \qquad (5.29)$$

For the situation without oversampling ($M_{\text{Tx}} = 1$), (5.7) leads to $R_{s,n-n'} = 2\sigma^2 \delta_{n,n'}$. From (5.14) and (5.15), we thus have

$$R^C_{s,n-n'} = 2 \sum_{\gamma=0}^{+\infty} P_{2\gamma+1} = 2P_1 + 2 \sum_{\gamma=1}^{+\infty} P_{2\gamma+1} \qquad (5.30)$$

for $n = n'$ and $R^C_{s,n-n'} = 0$ for $n' \neq n$; as a consequence,

$$\text{SIR}_k = \frac{P^C_1}{\sum_{\gamma=1}^{+\infty} P^C_{2\gamma+1}}, \qquad (5.31)$$

which is independent of k. It should be noted that, for $M_{\text{Tx}} > 1$ (i.e., when $N' > N$), SIR_k is a function of k, since $E[|D^C_k|^2]$ is also a function of k.

When N is high enough to validate a Gaussian approximation for the time-domain samples at the nonlinearity input, s_n, our modeling approach for the final frequency-domain and time-domain blocks is quite accurate, and D^C_k exhibit quasi-Gaussian characteristics for any k [AD10b, AD12b].

5.1.3 Numerical Results

In this subsection, we present a set of performance results concerning the signal processing schemes studied here. We consider OFDM schemes with $N = 256$ subcarriers, unless otherwise stated, and two types of clipping: envelope clipping and Cartesian clipping. For these clipping characteristics, it can easily be shown that S^C and I^C can be written in closed form. In fact, $S^C = |\alpha^C|^2 \sigma^2$, with α^C given by (B.17)

$$\alpha^C = 1 - 2Q\left(\frac{s_M}{\sigma}\right), \tag{5.32}$$

for Cartesian clipping and by (B.41)

$$\alpha^C = 1 - e^{-\frac{s_M^2}{2\sigma^2}} + \frac{\sqrt{2\pi} s_M}{2\sigma} Q\left(\frac{s_M}{\sigma}\right), \tag{5.33}$$

for polar clipping, with $Q(\cdot)$ denoting the well-known error function, defined as

$$Q(x) \triangleq \frac{1}{\sqrt{2\pi}} \int_x^{+\infty} e^{-\frac{u^2}{2}} \, du. \tag{5.34}$$

With respect to I^C, we have $I^C = P_{\text{out}}^C - S^C$, with P_{out}^C given by (B.18)

$$P_{\text{out}}^C = 2\sigma^2 \left(\frac{1}{2} - \frac{s_M}{\sigma\sqrt{2\pi}} e^{-\frac{s_M^2}{2\sigma^2}} - \left(1 - \frac{s_M^2}{\sigma^2}\right) Q\left(\frac{s_M}{\sigma}\right)\right), \tag{5.35}$$

for Cartesian clipping, and by (B.42)

$$P_{\text{out}}^C = \sigma^2 \left(1 - e^{-\frac{s_M^2}{2\sigma^2}}\right), \tag{5.36}$$

for polar clipping.

The scaling factor α^C is depicted in Figure 5.3, and the constant SIR when $M_{\text{Tx}} = 1$ is depicted in Figure 5.4. As expected, the values of α and SIR increase with s_M/σ. For very high values of s_M/σ, $\alpha \to 1$ and SIR$\to +\infty$.

By using some oversampling, i.e., $M_{\text{Tx}} > 1$, the value of SIR$_k$ increases for all subcarriers and becomes no longer constant, with the subcarriers at the central region of the spectrum having the lower value, as shown in Figure 5.5. In all cases, the values of SIR$_k$ with $M_{\text{Tx}} = 2$ are very close to the ones obtained with $M_{\text{Tx}} \to +\infty$. Of course, the set of multiplying coefficients $\{F_k; k = 0, 1, \ldots, N' - 1\}$, used to shape the Power Spectral Density (PSD) of the transmitted signals (as will be shown in the following), does not change the SIR$_k$ levels in all cases. The power distribution of the self-interference component within the subcarriers, required for the computation of the SIR$_k$ values, can be obtained from $R_{d,n}^C$, given by (5.17) or (5.18), as described in Subsection 5.1.2 (see also (5.27)). From Figures 5.4 and 5.5, it is clear that for a given s_M/σ (and, inherently, a given maximum output envelope), the polar clipping has higher SIR$_k$ values.

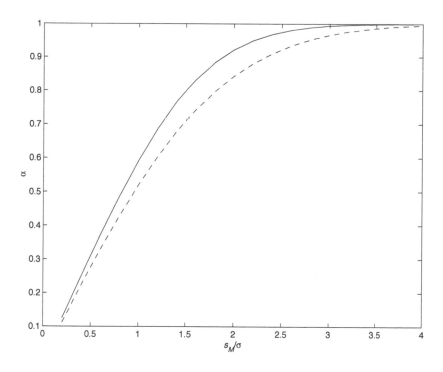

Figure 5.3: Scaling factor α^C for Cartesian (dashed line) and polar (solid line) clippings, as function of s_M/σ.

It should be noted that the statistical characterization of the transmitted blocks is independent from the number of subcarriers, provided that N is high enough to validate the Gaussian approximation. In this case, the statistical characterization is also independent from the adopted constellation on each subcarrier; however, since the self-interference noise behaves just as an additional Gaussian noise, the impact of a given SIR level changes with the adopted constellation. In the following, we will show that, when combined with a given OFDM transmission scheme, the signal processing schemes considered here can reduce the envelope fluctuation of the transmitted signals while keeping a high spectral efficiency.

As it was already mentioned, the statistical characterization of the transmitted blocks presented here is based on the Gaussian behavior of the samples at the nonlinearity input, s_n, which is valid when $N \to +\infty$. Since the exact analytical characterization with a small-to-moderate

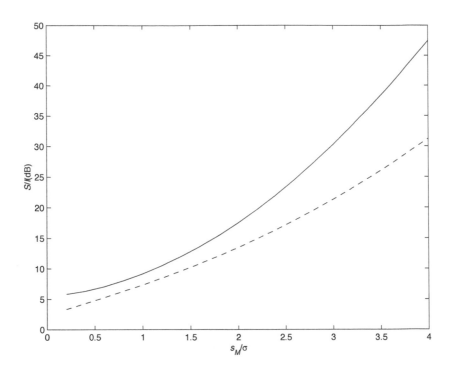

Figure 5.4: SIR levels when $M_{\mathrm{Tx}} = 1$, for Cartesian (dashed line) and polar (solid line) clippings, as functions of s_M/σ.

number of subcarriers is very complex (see [AD10b, AD12b] for details), it is important to know how accurate our method is for low values of N.

5.1.4 *Analytical BER Performance*

It should be mentioned that the proposed signal processing schemes only involve modifications at the transmitter side, being compatible with conventional OFDM receivers. The received signal is submitted to the receiver's filter, with impulse response $h_R(t)$, and then sampled at a rate $F_S = N''/T$, leading to blocks $\{y_n; n = 0, 1, \ldots, N'' - 1\}$ after removal of the guard period. A square-root raised-cosine receiver filtering is assumed (as well as for the transmit filtering), with roll-off $\rho \in \,]0, 1]$; it is also assumed that $|H_R(f)| = |H_T(f)| = 1$ in the frequency band, where $F_k \neq 0$. Next, a DFT operation leads to $\{Y_k; k = 0, 1, \ldots, N'' - 1\} = \mathrm{DFT}\,\{y_n; n = 0, 1, \ldots, N'' - 1\}$, where Y_k is concerned to the kth

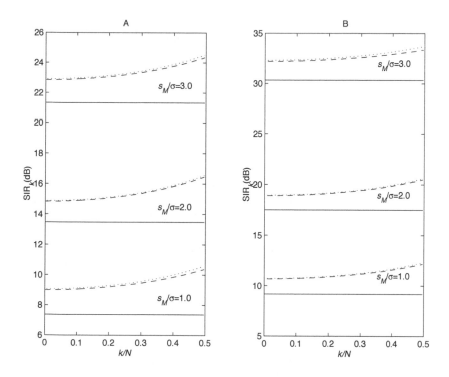

Figure 5.5: Evolution of SIR_k for Cartesian (A) and polar (B) clippings, when $M_{\mathrm{Tx}} = 1$ (solid line), 2 (dashed line) or $+\infty$ (dotted line).

subchannel. It is well-known that, thanks to the cyclic prefix,

$$Y_k = H_k S_k^{\mathrm{Tx}} + N_k \tag{5.37}$$

(when the guard interval is longer than the overall channel impulse response), where N_k and H_k denote the noise component and the overall channel frequency response, respectively, at the kth subchannel. It can also be shown that these frequency-domain noise samples, uncorrelated and Gaussian, are zero-mean, and the variance of their real and imaginary parts is given by $N_0(N'')^2/T$, when $N_0/2$ denotes the PSD of the white Gaussian noise at the receiver input.

Next, we will show how one can obtain the BER performance with the signal processing schemes considered here for a frequency-selective channel, characterized by the set of coefficients $\{H_k; k = 0, 1, \ldots, N'' - 1\}$, when conventional OFDM receivers are employed.

It was already shown that the frequency-domain block at the nonlinearity output can be decomposed into two uncorrelated components: a useful component, $\{\alpha F_k S_k; k = 0, 1, \ldots, N' - 1\}$, and a self-interference component $\{D_k F_k; k = 0, 1, \ldots, N' - 1\}$, where the D_k coefficients have an approximately Gaussian distribution. This means that, besides the channel noise component, the received frequency-domain block can also be decomposed into uncorrelated useful and self-interference components, since the frequency-selective channel affects in the same way the complex symbol to be transmitted and the corresponding self-interference $(H_k S_k^{\mathrm{Tx}} = \alpha H_k F_k S_k + H_k F_k D_k)$.

This decomposition of the frequency-domain blocks into useful and self-interference components involves two complementary degradation effects: on the one hand, just part of the transmitted power is useful; on the other hand, the self-interference component is added to the channel noise, leading to an additional degradation.

For an ideal, coherent receiver with perfect synchronization and channel estimation, the BER for the kth subchannel can be expressed as

$$P_{b,k} \approx \alpha_{M,k} Q\left(\sqrt{\beta_{M,k} \cdot \mathrm{ESNR}_k}\right), \qquad (5.38)$$

for $0 \geq k \geq N/2 - 1$ and $N'' - N/2 \geq k \geq N'' - 1$. In the equation above, $\alpha_{M,k}$ and $\beta_{M,k}$ are functions of the adopted constellation, $Q(\cdot)$ denotes the well-known error function given by (5.34), and ESNR_k denotes an 'equivalent signal-to-noise plus self-interference ratio' (ESNR). This ratio is given by

$$\mathrm{ESNR}_k = \frac{\alpha F_k H_k E[|S_k|^2]}{F_k H_k E[|D_k|^2] + E[|N_k|^2]}, \qquad (5.39)$$

where $E[|N_k|^2] = 2N_0(N')^2/T$ and the other expectations can be analytically computed as described above. Of course, the overall BER is then the average of the BER associated to each of the N in-band subchannels.

Asymptotically, when the channel noise effects become negligible, (5.38) takes the form

$$P_{b,k} \approx \alpha_{M,k} Q\left(\sqrt{\beta_{M,k} \cdot \mathrm{SIR}_k}\right), \qquad (5.40)$$

with SIR_k given by (5.29), which means that the nonlinear distortion should lead to an irreducible BER, depending on SIR_k only, for any given constellation.

As described above, when $M_{\mathrm{Tx}} = 1$, the SIR_k is independent of k and given by S^C/I^C, with S^C denoting the overall useful power and I^C denoting the overall self-interference power. This allows very simple BER computations, since we do not need to evaluate separately the powers of the different IMPs, $P_{2\gamma+1}^C$, required for the computation of $E[|D_k^C|^2]$.

5.1.5 Performance Results

In this section, we present a set of performance results concerning the basic signal processing schemes studied here. We consider an OFDM modulation scheme, with $N=256$ subcarriers and a square Quadrature Amplitude Modulation (QAM) constellation, with $M = 2^{2m}$ points, with a Gray mapping rule on each subcarrier (in fact, the performance of the signal processing schemes proposed here is almost independent of N, provided that N is high enough to allow the Gaussian approximation of the OFDM signals). The set of multiplying coefficients $\{F_k; k = 0, 1, \ldots, N'-1\}$ has a trapezoidal shape, with $F_k = 1$ for the N data subcarriers (in-band region), dropping linearly to 0 along the first $(N_1 - N)/2$ out-of-band subcarriers at both sides of the in-band region (this means that we have N_1 non-zero subcarriers). We consider the two clipping characteristics previously described: an ideal Cartesian clipping and an ideal envelope clipping. We assume in all cases a square-root raised-cosine shape with $\rho = 0.25$ and a one-sided bandwidth N''/T. It is also assumed that the transmitter employs a power amplifier that is quasi-linear within the range of variations of the input envelope.

Figures 5.6 and 5.7 are concerned to the bandwidth efficiency issues with, respectively, 'conventional OFDM' schemes and 'modified OFDM' schemes, by using a signal processing of the proposed class. A well-known PSD-related function was adopted in both cases: the so-called Fractional Out-of-Band Power (FOBP). The FOBP results in Figure 5.6 are helpful to evaluate the impact of the windowing choice within a conventional OFDM scheme: even when T_W is just a very small fraction of T, the spectrum becomes much more compact than with $T_W = 0$; on the other hand, the impact of $T_G/T \neq 0$ is rather small.

Figure 5.7 shows the FOBP when using the modified OFDM schemes proposed here, with $s_M/\sigma = 2.0$, for Cartesian and polar clippings and $T_G/T = 0.2$. Clearly, this clipping can lead to high out-of-band radiation levels. By using a frequency-domain filtering as reported above, with $N_1/N = 1$ (i.e., $F_k = 1$ for the N data subcarriers and 0 for the remaining $N' - N$ ones), we can reduce the out-of-band radiation to the levels of conventional OFDM; a further reduction in the out-of-band radiation levels is achieved if this frequency-domain filtering is combined with $T_W/T = 0.025$ (the use of $T_W/T = 0.025$ with $F_k = 1$ for any k practically does not provide any improvement in the FOBP when $s_M/\sigma = 2.0$). For other values of s_M/σ, we can still have out-of-band radiation levels similar to those of conventional OFDM schemes, provided that N_1/N is low enough.

In Figure 5.8 we show the envelope distribution of the samples s_n^{CF} for a polar clipping (similar results are obtained for Cartesian clipping).

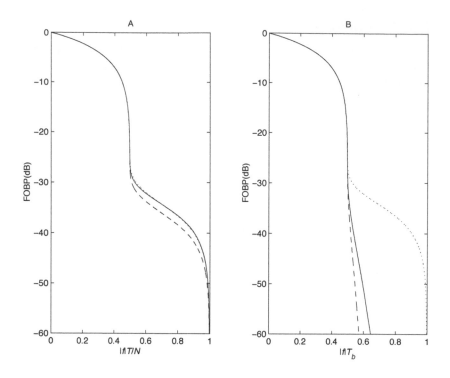

Figure 5.6: FOBP with conventional OFDM: (A) for $T_W = 0$ **and** $T_G/T = 0$ **(solid line), 0.2 (dotted line) or 0.5 (dashed line); (B) for** $T_G/T = 0.2$ **and** $T_W/T = 0$ **(dotted line), 0.025 (solid line) and 0.05 (dashed line).**

As expected, an overall reduction of several dB can be observed on the resulting PMEPR.

Figure 5.9 shows the PMEPR, defined according to (2.90) with $P = 10^{-3}$, when $N_1 = N$, $N_1 = 1.5N$ or, for the sake of comparisons, when we do not employ any frequency-domain filtering ($F_k = 1$ for any k), for different values of M_{Tx}. From this figure we can see that the PMEPR increases with s_M/σ and is slightly higher for $N_1 = N$ (i.e., when lower out-of-band radiation levels are intended). We can also see that the higher the oversampling factor, the lower the PMEPR; on the other hand, a high filtering effort leads to a regrowth of the PMEPR. Without frequency-domain filtering, we just have with $M_{Tx} = 4$ a PMEPR close to the one for $M_{Tx} \to +\infty$; however, under a strong

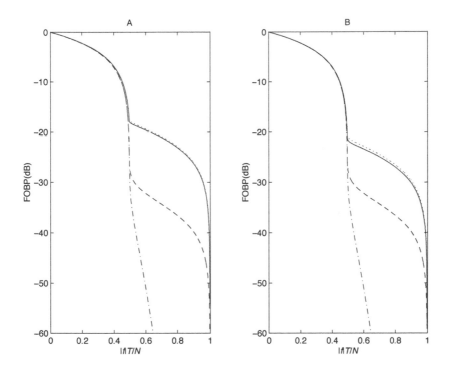

Figure 5.7: FOBP for Cartesian (A) and polar (B) clippings, when $N_1 = N$, $s_M/\sigma = 2.0$ and $T_G/T = 0.2$, for $T_W = 0$ (dashed line) or $T_W/T = 0.025$ (dash-dotted line). For the sake of comparisons, we include the FOBP when $F_k = 1$ for any k and $T_W/T = 0$ (dotted line) or 0.025 (solid line).

filtering effort, the PMEPR for $M_{Tx} = 2$ is already close to the PMEPR for $M_{Tx} \to +\infty$.

In Figure 5.10 we present the required E_b/N_0 to have BER $= 10^{-4}$ on an ideal Additive White Gaussian Noise (AWGN) channel, when $T_G = T_W = 0$ and we employ Cartesian and polar clippings (an ideal coherent receiver with perfect synchronization was assumed). As expected, the required E_b/N_0 decreases with s_M/σ in both cases. Clearly, the BER performances with $M_{Tx} = 2$ are already close to the corresponding performances with $M_{Tx} \to +\infty$.

Figure 5.11 shows the required $E_b^{(p)}/N_0$ to have BER $= 10^{-4}$ ($E_b^{(p)}/N_0 = E_b/N_0+$ PMEPR (dB)). This 'peak E_b/N_0' is appropriate for an overall comparison of the 'power efficiency' with the different

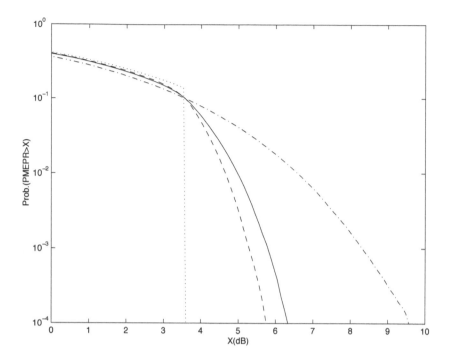

Figure 5.8: Envelope distribution for a polar clipping with $s_M/\sigma = 2.0$, with $N_1 = N$ (solid line), with $N_1 = 1.5N$ (dashed line), with $F_k = 1$ for any k (dotted line) and with conventional OFDM (dash-dotted line).

transmission alternatives, since it combines the 'detection efficiency' issue (through the required E_b/N_0) and the requirements on power amplification back-off (through the PMEPR). Clearly, the envelope clipping (polar clipping) has a better power efficiency than the I-Q clipping (Cartesian clipping). From Figure 5.11, we can see that the optimum values of s_M/σ for a Cartesian clipping are 2.6 for $M = 4$, 3.6 for $M = 16$ and 4.2 for $M = 64$; for a polar clipping, the optimum values of s_M/σ are 2.0 for $M = 4$, 2.6 for $M = 16$, and 3.2 for $M = 64$. In both cases, the optimum values of s_M/σ are almost independent of the filtering effort.

An important issue is the accuracy of the analytical approach for the performance evaluation presented here, which relies on both the quasi-Gaussian nature of OFDM signals with a high number of subcarriers, and the quasi-Gaussian characterization of the self-interference term at the subcarrier level. Our simulation results show that, whenever $N \geq 64$,

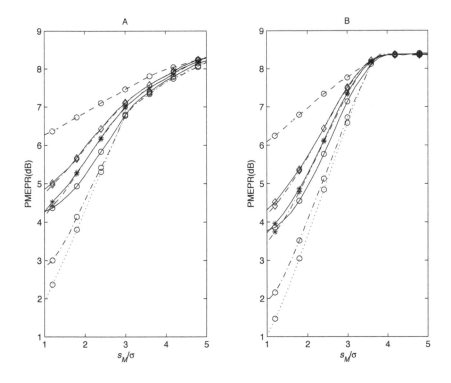

Figure 5.9: PMEPR for Cartesian (A) and polar (B) clippings, with $N_1 = N$ (◊), $N_1 = 1.5N$ (*) and for $F_k = 1$ for any k (o), when $M_{\mathbf{Tx}} = 1$ (dashed line), 2 (solid line), 4 (dash-dotted line) or $+\infty$ (dotted line).

the analytical approach for computation of the PSD of the transmitted signals is very accurate, regardless the type of clipping, the clipping level, the oversampling factor and the constellation size.

However, the accuracy of the analytical approach for computation of the BER performance depends on the number of subcarriers but also on the clipping level and the constellation size. For example, the theoretical and simulated BER performances of Figure 5.12, for a polar clipping when $M_{\mathrm{Tx}} = 2$ and $N = 256$, are in close agreement (similar conclusions could be taken for other oversampling factors and for a Cartesian clipping). For $N = 64$, the analytical approach is not accurate when we are close to the irreducible error floor associated to the self-interference: for small values of s_M/σ, the theoretical performance is slightly pessimistic; the analytical SIR_k is slightly lower than the 'true' SIR_k. For

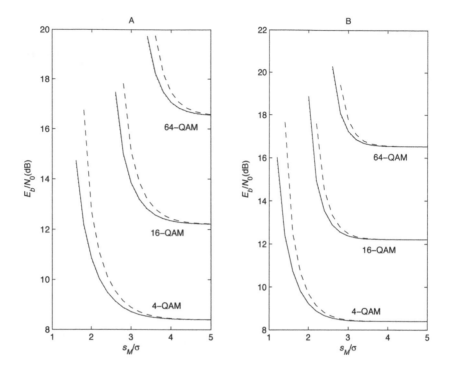

Figure 5.10: Required E_b/N_0 for BER= 10^{-4}, with Cartesian (A) and polar (B) clippings and $F_k = 1$ for any k, when $M_{\mathbf{Tx}} = 1$ (dashed line), 2 (solid line) and $+\infty$ (dotted line).

higher values of s_M/σ, the self-interference term at the subcarrier level can no longer be assumed to be Gaussian; however, we can still employ our analytical approach, since the corresponding self-interference levels are usually significantly below the channel noise levels.

5.1.6 Complementary Remarks

The clipping techniques described above allow significant reductions on the envelope fluctuations of multicarrier signals and, therefore, a more efficient power amplification. However, we still need to employ quasi-linear amplifiers. However, it is known that grossly nonlinear amplifiers are simpler, cheaper, have higher output power and have higher amplification efficiency. Therefore, it would be desirable to employ grossly nonlinear power amplifiers with multicarrier signals. A promising way of doing

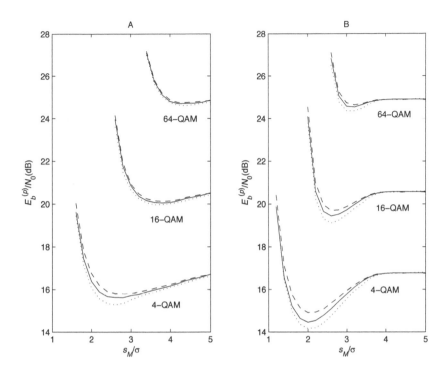

Figure 5.11: Required $E_b^{(p)}/N_0$ for BER $= 10^{-4}$, with Cartesian (A) and polar (B) clippings and $M_{\mathbf{Tx}} = 2$, for the following cases: $F_k = 1$ for any k (dotted line); $N_1 = N$ (dashed line); $N_1 = 1.5N$ (solid line).

this is to use two-branch, Linear Amplification with Nonlinear Components (LINC) transmitter structures [Cox74]. LINC techniques were proposed for OFDM signals [DG96b, DG01, DG08] and are in the basis of the more exotic CEPB-OFDM schemes (Constant-Envelope Paired-Burst OFDM) that allow a single grossly nonlinear power amplifier at the transmitter [DG96a, DG98, DG03a]. Moreover, CEPB-OFDM signals have intrinsic characteristics that help the carrier synchronization procedure [DG97]. The analytical methods described here can easily be extended to study LINC and CEPB-OFDM techniques.

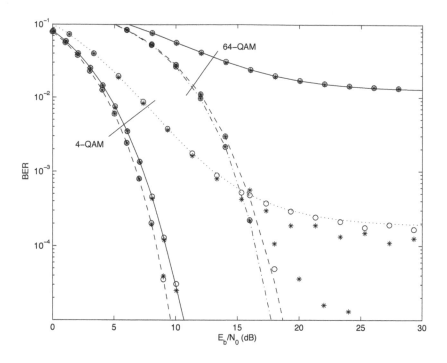

Figure 5.12: BER performance for 4 QAM (A) and 64 QAM constellations (B), when a polar clipping is employed with $s_M/\sigma = 1.0$ (dotted line), 2.0 (solid line), 3.0 (dashed line) and 4.0 (dash-dotted line) ((o) for $N = 256$ and (*) for $N = 64$).

5.2   Impact of Quantization Effects on Multicarrier Signals

This section is dedicated to the study of quantization effects on the complex envelope of multicarrier signals. These quantization effects occur at both the transmitter and the receiver and are associated to the numerical accuracy of the DFT computations.

The evaluation of quantization effects is a well-known problem in analog-to-digital conversion. The usual approach is to assume that uniformly distributed noise is added to the quantized signal [Gra90, WKL96]. However, this approach is not suitable if there are clipping effects (i.e., the 'saturation' of the quantizer is not a very rare event) or for non-uniform quantizers. A general theory for non-uniform quantizers can be found in [JN84]. An analytical approach for evaluating the impact of memoryless nonlinear devices in real-valued Gaussian signals was presented in [Dar03] and used for evaluating quantization and clipping effects at the multicarrier receiver. The impact of the oversampling factor on the SIR levels was also considered there.

The basic results from this section were published in [AD07b].

5.2.1 Quantization Effects on Multicarrier Modulated Signals

Figure 5.13 presents the transmission chain considered for the evaluation of quantization effects on multicarrier signals. Each DFT/IDFT operation is modeled as an ideal DFT/IDFT operation (i.e., with infinite precision), followed and preceded by quantization blocks, modeling numerical accuracy issues and clipping effects (see [THM72]).[1] However, as shown in Figure 5.13, we will ignore the quantization effects on the frequency-domain samples. This can be justified as follows:

■ The quantization effects in the frequency-domain samples (i.e, at the input of the 'ideal IDFT device' and the output of the 'ideal DFT device') have a local effect, restricted to a given subcarrier. On the other hand, since the quantization effects in the time-domain samples produce spectral widening (as it will be shown in the following), the resulting nonlinear distortion effects can affect all subcarriers.

■ The frequency-domain signals to be transmitted have lower dynamic range than the corresponding time-domain signals (e.g., belonging to a given Phase-Shift Keying (PSK) or QAM constellation, instead of having a quasi-Gaussian distribution).

■ The received samples are usually submitted to a decision device (which can be modeled as a suitable quantization characteristic).

As seen in Chapter 2, each of the N frequency-domain symbols to be transmitted is selected from a given constellation, according to the transmitted data, and an augmented block is formed by adding $N' - N$ idle subcarriers, i.e., with $S_k = 0$. The complex envelope of the bandpass signal is referred to the frequency $f_c = f_0 + \Delta N F$ (which corresponds to the zero frequency of the complex envelope), where f_0 is the central frequency of the bandpass signal and F denotes the subcarrier separation (see Figure 2.4). In conventional multicarrier implementations, the

---

[1]Basically, this means that the DFT operation is performed ideally (without numerical errors), and the effect of the numerical errors is simply included by ignoring the least reliable bits.

complex envelopes are referred to the central frequency of the spectrum, leading to $f_c = f_0$ and $\Delta N = 0$. It will be shown in the following that the selected ΔN can have a significant effect on the quantizers performance.

An 'ideal IDFT device' produces the time-domain block $\{s_n; n = 0, 1, \ldots, N'-1\} = \text{IDFT}\{S_k; k = 0, 1, \ldots, N'-1\}$, which can be regarded as a sampled version of the multicarrier burst, with an oversampling factor $M_{\text{Tx}} = N'/N$ (see (2.28)), followed by a quantization operation (which includes clipping effects), leading to the block of time-domain samples to be transmitted $\{s_n^Q; n = 0, 1, \ldots, N'-1\}$. The quantizer operates on complex-valued symbols and can be regarded as a Cartesian memoryless nonlinearity, which separately operates on the real and the imaginary parts of each complex sample x [DG01]. This is done in accordance with

$$s_n^Q = g_{Q1}(\text{Re}\{s_n\}) + jg_{Q1}(\text{Im}\{s_n\}), \tag{5.41}$$

where $g_{Q1}(x)$ denotes an appropriate odd function.

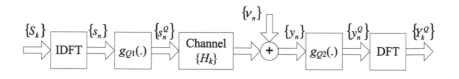

Figure 5.13: Multicarrier transmission chain with quantization blocks.

When the guard interval is longer than the length of the overall channel impulse response, the received time-domain samples, y_n, are such that $\{y_n; n = 0, 1, \ldots, N'-1\} = \text{IDFT}\{Y_k; k = 0, 1, \ldots, N'-1\}$, where the received symbol for the kth subcarrier is given by (2.36), with the quantized samples S_k^Q replacing the original samples S_k

$$Y_k = H_k S_k^Q + N_k, \tag{5.42}$$

with H_k and N_k denoting the corresponding channel frequency response and channel noise, respectively, and $\{S_k^Q; k = 0, 1, \ldots, N'-1\} = \text{DFT}\{s_n^Q; n = 0, 1, \ldots, N'-1\}$.

The samples y_n are quantized before the DFT operation, leading to the samples

$$y_n^Q = g_{Q2}(\text{Re}\{y_n\}) + jg_{Q2}(\text{Im}\{y_n\}), \tag{5.43}$$

also with an appropriate odd function $g_{Q2}(x)$. The DFT operation produces the frequency-domain block $\{Y_k^Q; k = 0, 1, \ldots, N' - 1\}$, which is used for the detection of the transmitted symbols.

Since we consider floating-point operations, we assume functions $g_{Q1}(x)$ and $g_{Q2}(x)$ take the form (3.5), i.e.,

$$g_{\text{quant}}(x) = \pm m(x) 2^{-e(x)} s_M \tag{5.44}$$

where s_M denotes the saturation level (see Figure 3.3).

5.2.2 Statistical Characterization of the Transmitted and Received Signals

When the number of subcarriers is high $(N \gg 1)$ the time-domain coefficients s_n can be approximately regarded as samples of a zero-mean complex Gaussian process (see Section 2.4). If $E[S_k S_{k'}^*] = G_{S,k} \delta_{k,k'}$, then $E[s_n] = 0$ and the autocorrelation of the time-domain samples s_n is given by (2.74)

$$E[s_n s_{n'}^*] = R_{s,n-n'} = \frac{1}{(N')^2} \sum_{k=0}^{N'-1} G_{S,k}\, e^{j2\pi \frac{k(n-n')}{N'}}, \tag{5.45}$$

$n, n' = 0, 1, \ldots, N' - 1$, i.e., with $\{R_{s,n}; n = 0, 1, \ldots, N' - 1\} = \frac{1}{N'}$ IDFT $\{G_{S,k}; k = 0, 1, \ldots, N' - 1\}$. The variance of both $\text{Re}\{s_n\}$ and $\text{Im}\{s_n\}$ is

$$\sigma^2 = \frac{1}{2} E[|s_n|^2] = \frac{1}{2} R_{s,0} = \frac{1}{2(N')^2} \sum_{k=0}^{N'-1} G_{S,k}. \tag{5.46}$$

We will now take advantage of the mentioned quasi-Gaussian nature of the samples s_n for obtaining the statistical characterization of the transmitted blocks. The output of a memoryless nonlinear device with a Gaussian input can be written as the sum of two uncorrelated components: a useful one, proportional to the input, and a self-interference one (see (3.30)). Since the real and imaginary parts of the quantizer input are submitted to identical memoryless nonlinearities, its output can be written as

$$s_n^Q = \alpha^{Q1} s_n + d_n^{Q1}, \tag{5.47}$$

where $E[s_n d_{n'}^{Q1*}] = 0$ and α^{Q1} is given by (3.31)

$$\alpha^{Q1} = \frac{E[x g_{Q1}(x)]}{E[x^2]} = \frac{1}{\sqrt{2\pi}\sigma^3} \int_{-\infty}^{+\infty} x g_{Q1}(x)\, e^{-\frac{x^2}{2\sigma^2}}\, dx. \tag{5.48}$$

The average power of the useful component is $S^{Q1} = |\alpha^{Q1}|^2 \sigma^2$, and the average power of the self-interference component is given by $I^{Q1} = P_{\text{out}}^{Q1} - S^{Q1}$, where P_{out}^{Q1} denotes the average power of the signal at the nonlinearity output, given by (3.34)

$$P_{\text{out}}^{Q1} = E[g_{Q1}^2(x)] = \frac{1}{\sqrt{2\pi}\sigma} \int_{-\infty}^{+\infty} g_{Q1}^2(x)\, e^{-\frac{x^2}{2\sigma^2}}\, dx. \qquad (5.49)$$

In Chapter 3, it was shown that the autocorrelation of the output samples, $R_{s,n-n'}^{Q} = E[s_n^Q s_{n'}^{Q*}]$, can be expressed as a function of the autocorrelation of the input samples as (see (3.67))

$$R_{s,n-n'}^{Q} = 2 \sum_{\gamma=0}^{+\infty} P_{2\gamma+1}^{Q1} \frac{(\text{Re}\{R_{s,n-n'}\})^{2\gamma+1} + j(\text{Im}\{R_{s,n-n'}\})^{2\gamma+1}}{(R_{s,0})^{2\gamma+1}}, \qquad (5.50)$$

where $R_{s,n-n'}$ is given by (5.45) and the coefficient $P_{2\gamma+1}^{Q1}$, denoting the total power associated to the IMP of order $2\gamma + 1$, is computed from (3.51). Since

$$R_{s,n-n'}^{Q} = |\alpha^{Q1}|^2 R_{s,n-n'} + E[d_n^{Q1} d_{n'}^{Q1*}], \qquad (5.51)$$

it can be easily recognized that $P_1^{Q1} = |\alpha^{Q1}|^2 \sigma^2$ and

$$\begin{aligned}
R_{d,n-n'}^{Q1} &= E[d_n^{Q1} d_{n'}^{Q1*}] \\
&= 2 \sum_{\gamma=1}^{\infty} P_{2\gamma+1}^{Q1} \frac{(\text{Re}\{R_{s,n-n'}\})^{2\gamma+1} + j(\text{Im}\{R_{s,n-n'}\})^{2\gamma+1}}{(R_{s,0})^{2\gamma+1}}.
\end{aligned}$$
$$\qquad (5.52)$$

Having in mind (5.47), the frequency-domain block $\{S_k^Q; k = 0, 1, \ldots, N' - 1\} = \text{DFT}\, \{s_n^Q; n = 0, 1, \ldots, N' - 1\}$ can obviously be decomposed into useful and self-interference components, i.e.,

$$S_k^Q = \alpha^{Q1} S_k + D_k^{Q1}, \qquad (5.53)$$

where $\{D_k^{Q1}; k = 0, 1, \ldots, N' - 1\}$ denotes the DFT of $\{d_n^{Q1}; n = 0, 1, \ldots, N' - 1\}$.

Since $g_{Q1}(x)$ is an odd function of x, it can be shown that $E[d_n^{Q1}] = 0$, leading to $E[D_k^{Q1}] = 0$. Moreover, as in (4.56),

$$E[D_k^{Q1} D_{k'}^{Q1*}] = \sum_{n=0}^{N'-1} \sum_{n'=0}^{N'-1} E[d_n^{Q1} d_{n'}^{Q1*}]\, e^{-j2\pi \frac{kn - k'n'}{N'}} = N' G_{D,k}^{Q1} \delta_{k,k'},$$
$$\qquad (5.54)$$

$k, k' = 0, 1, \ldots, N' - 1$, where $\{G_{D,k}^{Q1}; k = 0, 1, \ldots, N' - 1\}$ denotes the DFT of the block $\{R_{d,n}^{Q1}; n = 0, 1, \ldots, N' - 1\}$. This means that the self-interference components associated to different subcarriers are uncorrelated. Similarly,

$$E[S_k^Q S_{k'}^{Q*}] = N' G_{S,k}^Q \delta_{k,k'}, \qquad (5.55)$$

where $\{G_{S,k}^Q; k = 0, 1, \ldots, N' - 1\}$ denotes the DFT of $\{R_{s,n}^Q; n = 0, 1, \ldots, N' - 1\}$, given by (5.50), and

$$G_{S,k}^Q = |\alpha^{Q1}|^2 G_{S,k} + G_{D,k}^{Q1}. \qquad (5.56)$$

Let us consider now the transmission of the multicarrier signal over a time-dispersive channel. From (5.42), we have

$$y_n = s_n^Q * h_n + \nu_n = (\alpha^{Q1} s_n + d_n^{Q1}) * h_n + \nu_n$$

$$= \sum_{n'=0}^{N'-1} (\alpha^{Q1} s_{n-n'} + d_{n-n'}^{Q1}) h_{n'} + \nu_n, \qquad (5.57)$$

where ν_n represent the time-domain noise samples and $\{h_n; n = 0, 1, \ldots, N'-1\} = \text{IDFT} \{H_k; k = 0, 1, \ldots, N'-1\}$ is the channel impulse response of the equivalent discrete channel (as usual, it is assumed that $h_n = 0$ for $n > N_g$, with N_g denoting the number of samples in the guard interval (see Subsection 2.2.2)). Since the self-interference component d_n^{Q1} is not Gaussian, y_n is not Gaussian in the general case. However, if the channel has a large number of multipath components (as in severely time-dispersive channels with rich multipath propagation), h_n has a large number of non-zero terms, and the samples y_n can still be regarded as samples of a zero-mean Gaussian process[2]. The corresponding autocorrelation is $E[y_n y_{n'}^*] = R_{y,n-n'}$, where $\{R_{y,n}; n = 0, 1, \ldots, N' - 1\} = \frac{1}{N'}$ IDFT $\{G_{Y,k}; k = 0, 1, \ldots, N' - 1\}$, with

$$G_{Y,k} = E[|Y_k|^2] = |H_k|^2 G_{S,k}^Q + E[|N_k|^2]. \qquad (5.58)$$

This means that the samples y_n^Q can be decomposed into uncorrelated useful and self-interference components as in (5.47), i.e.,

$$y_n^Q = \alpha^{Q2} y_n + d_n^{Q2}, \qquad (5.59)$$

with α^{Q2} given by (3.31). The corresponding autocorrelation $R_{y,n-n'}^Q =$

---

[2]In fact, since the power of the self-interference component is typically much lower than the power of the useful component, the Gaussian approximation of the received samples can be very accurate, even for non-dispersive channels.

$E[y_n^Q y_{n'}^{Q*}]$ can be written as in (5.50), i.e.,

$$R_{y,n-n'}^Q = 2\sum_{\gamma=0}^{+\infty} P_{2\gamma+1}^{Q2} \frac{(\mathrm{Re}\{R_{y,n-n'}\})^{2\gamma+1} + j(\mathrm{Im}\{R_{y,n-n'}\})^{2\gamma+1}}{(R_{y,0})^{2\gamma+1}}.$$

(5.60)

Once again, the coefficient $P_{2\gamma+1}^{Q2}$ denotes the total power associated to the IMP of order $2\gamma + 1$ and can be computed from (3.51). The corresponding frequency-domain samples Y_k^Q can also be decomposed into useful and self-interference components,

$$Y_k^Q = \alpha^{Q2} Y_k + D_k^{Q2},$$

(5.61)

and $E[Y_k^Q Y_{k'}^{Q*}] = N' G_{Y,k}^Q \delta_{k,k'}$. Using (5.42) and (5.53), the received frequency-domain samples can be written as

$$Y_k^Q = \alpha^{Q1} \alpha^{Q2} H_k S_k + \alpha^{Q2} H_k D_k^{Q1} + \alpha^{Q2} N_k + D_k^{Q2},$$

(5.62)

hence, one can calculate an ESNR for each subcarrier, given by

$$\mathrm{ESNR}_k = \frac{|\alpha^{Q1} \alpha^{Q2} H_k|^2 E[|S_k|^2]}{|\alpha^{Q2} H_k|^2 E[|D_k^{Q1}|^2] + |\alpha^{Q2}|^2 E[|N_k|^2] + E[|D_k^{Q2}|^2]}.$$

(5.63)

It was observed that, when the number of active subcarriers is high enough to validate the Gaussian approximation for the time-domain samples at the nonlinearity input (say $N \geq 64$), our modeling approach is quite accurate. Moreover, under this 'high number of subcarriers' assumption, the self-interference terms D_k^{Q1} and D_k^{Q2} typically exhibit quasi-Gaussian characteristics for any k, although, as stated above, the self-interference components d_n^{Q1} and d_n^{Q2} are obviously not Gaussian [DG01, DG04]. The Gaussian approximation of the frequency-domain self-interference terms, D_k^{Q1} and D_k^{Q2}, is very accurate, unless very mild nonlinear characteristics are considered (in that case, almost all of the self-interference samples in the time-domain are zero and the corresponding frequency-domain samples are no longer Gaussian [BSGS02]).

Quantization characteristics are strongly nonlinear, hence the simplified computation of the power series in (5.50) and (5.60) presented in Section 5.1.2 can be applied.

5.2.3 Performance Results

In this section, we present a set of performance results concerning the quantization effects on multicarrier signals. We consider a multicarrier modulation with $N = 256$ active subcarriers (similar results were obtained for other values of N, provided that $N \gg 1$), with the same assigned power, and the quantization/clipping characteristics described in Section 3.1. Computations use $\gamma_{\max} = 40$.

Figure 5.14 shows the impact of the normalized saturation level, s_M/σ on the signal-to-interference ratio of the transmitted signals SIR_{Tx}, defined as the average over the in-band subcarriers of

$$\text{SIR}_k = \frac{|\alpha^{Q1}|^2 E[|S_k|^2]}{E[|D_k^{Q1}|^2]}, \tag{5.64}$$

for an uniform quantizer ($N_e = 0$), when $\Delta N = 0$ (i.e., when the complex envelope of the multicarrier signal is referred to the central frequency of the spectrum) and $M_{\text{Tx}} = 1$ (i.e., there is no oversampling). Clearly, there is an optimum value of s_M/σ that increases with the number of quantization bits. For small values of s_M/σ, the quantization noise is mainly a consequence of saturation effects, and SIR_{Tx} is almost independent of the number of quantization bits; for larger values of s_M/σ, we almost do not have saturation effects, and there is an improvement of 6 dB on SIR_{Tx} for each additional quantization bit (naturally, the larger s_M/σ, the higher the 'step-size' of the quantizer, leading to worse SIR_{Tx} levels).

Figure 5.15 shows the impact of the oversampling factor M_{Tx} on SIR_{Tx}, for an uniform quantizer. From this figure, we can observe improvements on the SIR_{Tx} levels when we increase the oversampling factor M_{Tx}, especially for moderate and high values of s_M/σ. These improvements are a consequence of the decreased aliasing effects in the in-band region when we increase the oversampling factor.

It should be pointed out that the behaviour of the SIR at the transmitter depicted in Figures 5.14 and 5.15 is similar to the behaviour of the SIR at the receiver when the impact of the channel is not considered (as in [Dar03]).

In Figure 5.16, we consider a non-uniform quantizer (i.e., $N_e \neq 0$). Although the non-uniform quantizers have good SIR_{Tx} levels for a wider range of saturation levels, for the optimum value of s_M/σ, the performance of a nonlinear quantizer is similar, or even better.

Figure 5.17 shows the gains on the SIR_{Tx} levels associated with $\Delta N \neq 0$. By adopting $\Delta N \approx N/2$ (i.e., by referring the complex envelope of the multicarrier signal to a frequency on the edge of the 'useful' band), these gains are approximately 1 dB for $M_{\text{Tx}} = 4$ and 2 dB for $M_{\text{Tx}} = 8$. The SIR gains when $\Delta N \neq 0$ can be explained in the following way: although the total self-interference power is independent of ΔN, when M_{Tx} is high enough, the adoption of $\Delta N \neq 0$ can lead to a decrease on the in-band self-interference levels and an increase on the out-of-band radiation levels (see Figure 5.18), leading to higher SIR values.

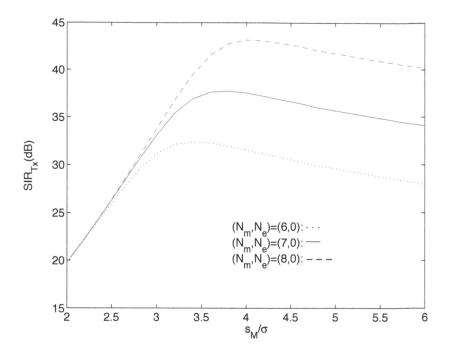

Figure 5.14: Impact of s_M/σ on SIR_{Tx}, when $\Delta N = 0$ and $M_{\text{Tx}} = 1$.

Let us consider now a severely frequency-selective channel (the typical situation for multicarrier transmission). At the detection level, the self-interference term associated with $g_{Q1}(\cdot)$ is $E[|D_k^{Q1} H_k|^2]$, which follows the evolution of the channel amplitude response. However, the self-interference term associated to $g_{Q2}(\cdot)$, $E[|D_k^{Q2}|^2]$, is almost independent of $|H_k|^2$. This means that the latter can have a significant effect on ESNR_k for the frequencies corresponding to deep fades, leading to higher quantization requirements at the receiver (for this reason, a higher number of quantization bits is assumed at the receiver). This behaviour is depicted in Figure 5.19, where we present the evolution of the analytical values of $E[|D_k^{Q2}|^2]$ when a severe time-dispersive channel is considered. For the sake of comparison, we also include values of $E[|D_k^{Q2}|^2]$ obtained by simulation, slightly better, but very close to the theoretical ones, with differences below 0.2 dB. The corresponding values of ESNR_k are depicted in Figure 5.20, together with the ESNR_k values for an ideal AWGN channel (for the same quantization characteristics and SNR). Since ESNR_k closely follows the evolution of $|H_k|^2$ (see Figure 5.19), it

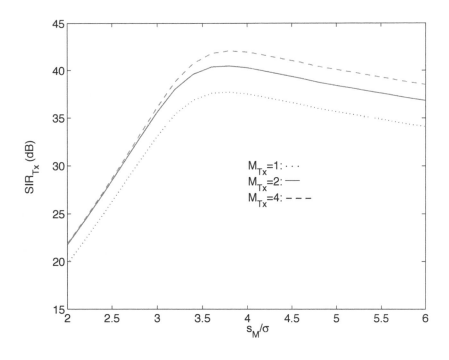

Figure 5.15: Impact of s_M/σ on $\mathrm{SIR_{Tx}}$, when $\Delta N = 0$ and $(N_m, N_e) = (7, 0)$.

can have large fluctuations for frequency selective channels, with very poor values for frequencies in deep fades (this in not the case of the ideal AWGN channel, or flat fading channels, where ESNR_k has only minor fluctuations in the in-band region).

5.3   Impact of Quantization Effects on Software Radio Signals

In this section, we present an analytical tool for evaluating the quantization requirements within the ADC used in software radio architectures. For this purpose, we take advantage of the Gaussian behaviour of the multi-band/multi-user signal at the input of the wideband ADC. This characterization is then used for the performance evaluation and optimization of different quantization characteristics.

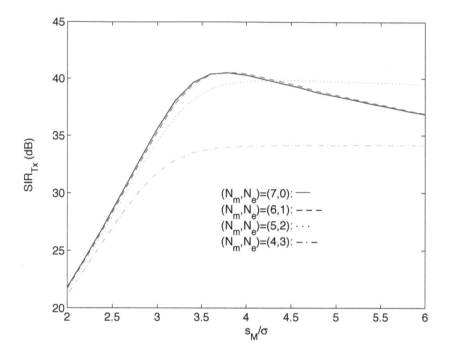

Figure 5.16: Impact of s_M/σ on SIR$_{\text{Tx}}$, for non-uniform quantizers, when $\Delta N = 0$ and $M_{\text{Tx}} = 2$.

The basic results from this section were published in [AD07a].

5.3.1 ADC Quantization in Software Radio Architectures

The software radio architecture considered in this section is similar to the one depicted in Figure 2.19, with samples z_n^Q replacing z_n. As seen in Section 2.3, we have P channels, each one with a bandpass signal centered on the frequency f_p, $p = 1, 2, \ldots, P$. The overall signal at the input of the receiver is given by (2.55). After the front-end filter, we obtain the signal $z(t)$, whose complex envelope referred to the frequency f_c is given by (2.57). As mentioned before, it is assumed that $f_c = f_0 + \Delta f$, i.e., the complex envelopes are referred to a frequency, which is shifted from the usual reference frequency by a factor $\Delta f = \Delta N F$ (see

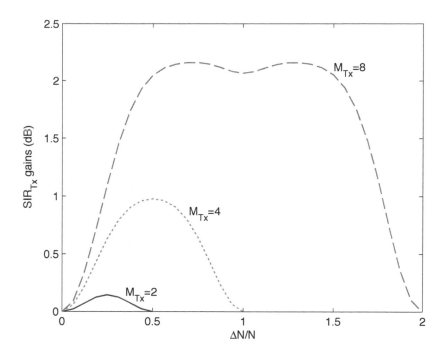

Figure 5.17: Gains on SIR$_{\text{Tx}}$ levels when $\Delta N \neq 0$ for $s_M/\sigma = 3.8$ and $(N_m, N_e) = (7, 0)$.

Figure 2.4). In the following, it will be shown that we can improve the ADC performance through an appropriate selection of f_c.

After down-conversion, $\tilde{z}(t)$ is sampled, leading to the samples z_n, and quantized. The quantizer operates on the complex-valued symbols z_n and can be regarded as a Cartesian memoryless nonlinearity operating separately on the real and the imaginary parts of each complex sample. This is done in accordance with (see Figure 3.8)

$$z_n^Q = g_{\text{quant}}(\text{Re}\{z_n\}) + j \, g_{\text{quant}}(\text{Im}\{z_n\}), \tag{5.65}$$

where $g_{\text{quant}}(x)$ denotes an appropriate quantization characteristic and is given by (3.5).

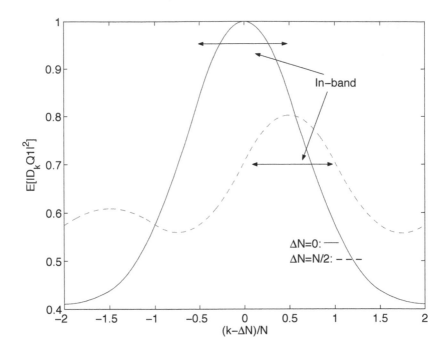

Figure 5.18: Normalized values of $E[|D_k^{Q1}|^2]$ **for** $\Delta N = 0$ **or** $\Delta N = 1/2$, **when** $s_M/\sigma = 3.8$, $(N_m, N_e) = (7, 0)$ **and** $M_{\mathbf{Tx}} = 4$.

5.3.2 Statistical Characterization of the Quantized Signals

When the number of channels is high or each channel has severe time dispersive effects, with rich multipath propagation, then $\tilde{z}(t)$ can be approximately regarded a zero-mean complex Gaussian process with PSD given by (2.73), autocorrelation given by (2.84), and the variance of both real and imaginary parts of $\tilde{z}(t)$ denoted by $\sigma^2 = R_{\tilde{z}}(0)/2$.

In this case, since the real and imaginary parts of the z_n^Q are submitted to two identical memoryless nonlinear operations, we have

$$z_n^Q = \alpha z_n + d_n, \tag{5.66}$$

with $E[z_n d_{n'}^*] = 0$, i.e., z_n^Q can be written as the sum of uncorrelated useful and self-interference components, and the analysis presented in Section 5.2 can be applied with samples z_n replacing samples s_n.

By summing the useful and self-interference frequency-domain sam-

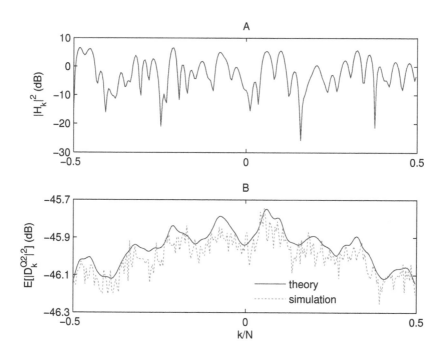

Figure 5.19: Evolution of $|H_k|^2$ (A) and theoretical and simulated values of $E[|D_k^{Q2}|^2]$ (B), when the channel SNR is 40 dB, $g_{Q1}(\cdot)$ is characterized by $(N_m, N_e) = (7, 0)$ and $s_M/\sigma = 3.8$ and $g_{Q2}(\cdot)$ is characterized by $(N_m, N_e) = (8, 0)$ and $s_M/\sigma = 4.0$.

ples for a given channel, we can obtain the SIR for each channel, SIR_p, $p = 1, 2, \ldots, P$.[3]

5.3.3 Performance Results

In this section, we present a set of performance results concerning the quantization effects on software radio architectures. We consider a wideband ADC with $P = 16$ channels and the quantization characteristics described in Subsection 5.3.1.

We will first assume that $\Delta f = 0$, i.e., the complex envelope of $y(t)$ is referred to the central frequency of its PSD ($f_c = f_0$). Figure 5.21 shows

---

[3]It is assumed that N is such that we have a high number of frequency-domain samples for each channel.

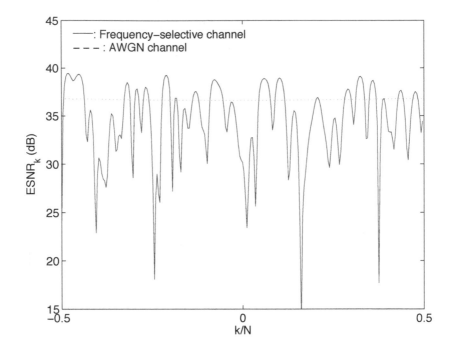

Figure 5.20: Evolution of ESNR$_k$ for the frequency-selective channel of Figure 5.19.A, as well as an ideal AWGN channel.

the impact of the normalized saturation level, s_M/σ, and the number of quantization bits N_m and N_e on the average value of SIR$_p$. For the sake of simplicity, it is assumed that $z(t)$ has a constant PSD in the inband region. From this figure, it is clear that there is an optimum value of s_M/σ for given N_e and N_m. We can also observe SIR improvements when we increase the oversampling factor.

Figure 5.22 shows the impact of Δf on the average value of SIR$_p$ for a uniform quantizer (i.e., $N_e = 0$) with $N_m = 7$. Clearly, the adequate selection of Δf allows an improvement on the SIR levels, when comparing with the case where $\Delta f = 0$: by adopting $\Delta f \approx B/2$ (i.e., by referring the complex envelope of $y(t)$ to a frequency on the edge of the 'useful' band), these improvements are approximately 1dB or 2dB for an oversampling factor of 4 or 8, respectively. The SIR improvements when $\Delta f \neq 0$ can be explained in the following way: although the total self-interference power is independent of Δf, when the oversampling factor is high enough, the adoption of $\Delta f \neq 0$ can lead to a decrease on

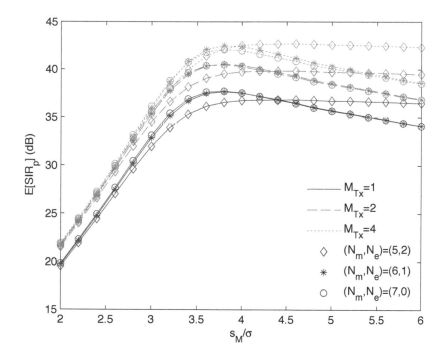

Figure 5.21: Impact of s_M/σ on the average value of SIR_p.

the in-band self-interference levels and an increase on the out-of-band radiation levels, as depicted in Figure 5.23, yielding higher SIR values.

Let us consider now different power distributions on each channel (see Figure 5.24.A). From Figure 5.24.B, it is clear that the PSD of $y(t)$ has just a minor impact on the PSD of the self-interference component, with differences below 1dB, even for severe time-dispersive channels (this behaviour was observed regardless of the oversampling factor, the adopted Δf and the type of quantizer). Since the self-interference component has an almost uniform PSD, the SIR levels are lower for the channels with a lower power.

It should be noted that our analytical results were validated by simulation, with differences below 0.5 dB. It should also be noted that the SIR gains with the oversampling factor and the appropriate selection of Δf are valid regardless of the quantization characteristics, even for a large number of quantization levels.

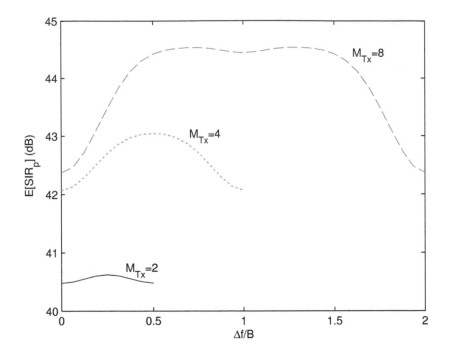

Figure 5.22: Impact of Δf on the SIR levels for $\Delta f = B/2$, $s_M/\sigma = 3.8$ and $(N_m, N_e) = (7, 0)$.

5.4 Loading Techniques with Nonlinear Distortion Effects

In this section, we study the impact of nonlinear distortion effects on adaptive multicarrier systems. An analytical statistical characterization of the transmitted signals is included and used to evaluate loading algorithms in the presence of strong nonlinear distortion effects and to redefine loading algorithms taking into account nonlinear distortion issues.

The basic results from this section were published in [AD09, AD12a].

5.4.1 Nonlinear Signal Processing for Multicarrier Signals

We consider nonlinear signal processing schemes, which operate on a sampled version of the multicarrier signal. The basic transmitter struc-

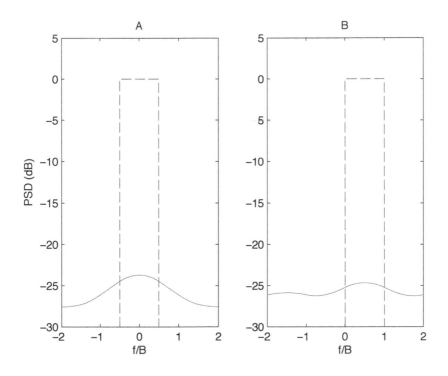

Figure 5.23: PSD of the useful component (dashed line) and the self-interference component (solid line) for $\Delta f = 0$ (A) or $\Delta f = B/2$ (B), with $M_{\mathbf{Tx}} = 4$, $s_M/\sigma = 3.8$ and $(N_m, N_e) = (4, 0)$.

ture considered in this section is depicted in Figure 5.25. This structure is similar to the clipping and filtering techniques proposed in [DG04, DG08] for reducing the PMEPR of the transmitted multicarrier signals while maintaining the spectral efficiency of conventional multicarrier schemes. As seen in Subsection 2.2.1, each of the N frequency-domain symbols to be transmitted, $\{\tilde{S}_k; k = 0, 1, \ldots, N-1\}$, is selected from a given constellation, according to the transmitted data, and an augmented block is formed by adding $N' - N$ idle subcarriers, i.e., with $S_k = 0$. An 'ideal IDFT device' produces the time-domain block $\{s_n; n = 0, 1, \ldots, N'-1\} = \text{IDFT}\{S_k; k = 0, 1, \ldots, N'-1\}$, which can be regarded as a sampled version of the multicarrier burst, with an oversampling factor $M_{\text{Tx}} = N'/N$. The statistical characterization of the augmented block was presented in Section 2.4.

The nonlinear device $g(\cdot)$ is modeled as a polar memoryless nonlinearity operating on an oversampled version of the multicarrier signal,

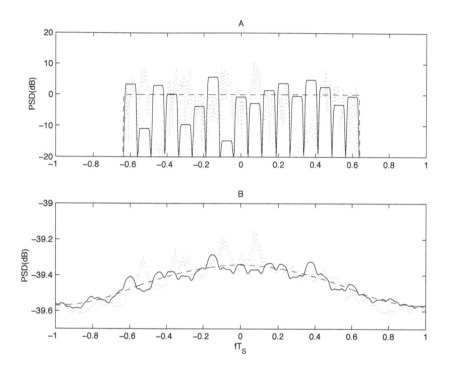

Figure 5.24: PSD of the useful component (A) and the self-interference component (B) of the quantized signal, when $(N_m, N_e) = (7,0)$ and $s_M/\sigma = 4.0$, for the following cases: $y(t)$ with uniform PSD and an AWGN channel (dashed line), $y(t)$ with non-uniform PSD and an AWGN channel (solid line), $y(t)$ with non-uniform PSD and frequency-selective channels (dotted line).

leading to the block of time-domain samples $\{s_n^C; n = 0, 1, \ldots, N' - 1\}$, with $s_n^C = g(s_n) = f(R)\, e^{j\,\arg(s_n)}$, where $R = |s_n|$. A DFT brings the nonlinearly modified block to the frequency domain, where an optional frequency-domain filtering procedure, through the set of multiplying coefficients $\{F_k; k = 0, 1, \ldots, N' - 1\}$, is considered to reduce the out-of-band radiation levels inherent to the nonlinear operation [DG04, DG08] (some of these coefficients are equal to zero in the out-of-band region). Since we consider a polar memoryless device, the approach presented in Section 4.2 can be used for the statistical characterization of the samples s_n^C. Hence, the nonlinearly distorted frequency-domain samples can be written as a sum of useful and nonlinear self-interference components,

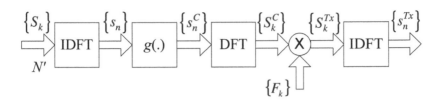

Figure 5.25: Detail of the transmitter structure with nonlinear signal processing.

as in (5.26), i.e.,

$$S_k^C = \alpha S_k + D_k, \tag{5.67}$$

with α given by (4.46). In Section 4.2, it was also shown that the autocorrelation of the output samples can be expressed as a function of the autocorrelation of the input samples as in (4.48), and that the PSD of the transmitted samples can be obtained from (4.58).

Having in mind (5.67) and the signal processing chain in Figure 5.25, the frequency-domain block $\{S_k^{Tx} = S_k^C F_k; k = 0, 1, \ldots, N' - 1\}$ can obviously be decomposed into useful and nonlinear self-interference components:

$$S_k^{Tx} = \alpha S_k F_k + D_k F_k, \tag{5.68}$$

where $\{D_k; k = 0, 1, \ldots, N' - 1\} = \text{DFT } \{d_n; n = 0, 1, \ldots, N' - 1\}$. Clearly, $E[D_k] = 0$ and, from (4.56), $E[D_k D_{k'}^*] = N' G_{D,k} \delta_{k,k'}$. Moreover, since D_k exhibits quasi-Gaussian characteristics for any k, provided that the number of subcarriers is high enough, $E[S_k^{Tx} S_{k'}^{Tx*}] = 0$ for $k \neq k'$, and $E[|S_k^{Tx}|^2] = |F_k|^2 E[|S_k^C|^2] = N'|F_k|^2 G_{S,k}^C$.

Let us now consider the transmission of the multicarrier signal over a time-dispersive channel. When the guard interval is longer than the length of the overall channel impulse response, the received time-domain samples y_n are such that $\{y_n; n = 0, 1, \ldots, N' - 1\} = \text{IDFT } \{Y_k; k = 0, 1, \ldots, N' - 1\}$, where the received symbol for the kth subcarrier is given by (2.36)

$$Y_k = H_k S_k^{Tx} + N_k, \tag{5.69}$$

with $\{S_k^{Tx} = F_k S_k^C; k = 0, 1, \ldots, N' - 1\} = \text{DFT } \{s_n^{Tx}; n = 0, 1, \ldots, N' - 1\}$. From (5.67) we can write

$$Y_k = \alpha F_k H_k S_k + F_k H_k D_k + N_k, \tag{5.70}$$

and an ESNR can be calculated for each subcarrier, given by

$$\text{ESNR}_k = \frac{|\alpha|^2 |F_k|^2 |H_k|^2 E[|S_k|^2]}{|F_k|^2 |H_k|^2 E[|D_k|^2] + E[|N_k|^2]}. \tag{5.71}$$

5.4.2   Loading Techniques

There are several bit loading algorithms designed to achieve spectrally-efficient communications proposed in the literature [HH87,LC97,FH96]. In this work, we use the algorithm presented in [LC97], but the analytical approach described in this work can easily be extended to other algorithms.

The algorithm's main objective is to maximize the data rate of the system, given a certain margin γ_m, which is kept fixed. This is carried out by assigning the number of bits to the various subcarriers and distributing the available energy among them, according to its signal-to-noise ratio. We assume the receiver performs coherent detection with perfect synchronization and channel estimation.

5.4.2.1   Single Carrier Analysis

It is well-known that the symbol energy for a square QAM constellation with 2^{2m} points is given by

$$\varepsilon = \frac{2^{2m} - 1}{6} d^2 \tag{5.72}$$

and that the Symbol Error Rate (SER) for moderate to high SNR regimes can be approximated by

$$P_s \approx 4 \left(1 - \frac{1}{2^m} \right) Q(\eta), \tag{5.73}$$

where $Q(\cdot)$ represents the well-known error function given by (5.34) and

$$\eta = \frac{d|H|}{2\sigma_N}, \tag{5.74}$$

with d representing the minimum Euclidean distance between constellation points, $|H|$ the channel gain and σ_N^2 the variance of the noise. Let SNR denote the channel output signal-to-noise ratio that is given by

$$\text{SNR} = \frac{\varepsilon|H|^2}{2\sigma_N^2}, \tag{5.75}$$

where ε denotes the symbol energy. Using (5.72), we can write

$$\eta = \sqrt{\beta_{\text{square}}\text{SNR}} \tag{5.76}$$

with

$$\beta_{\text{square}} \triangleq \frac{3}{2^{2m} - 1}. \tag{5.77}$$

Using (5.72) and (5.75), the transmitted number of bits can be written as

$$b = 2m = \log_2\left(1 + \frac{6\varepsilon}{d^2}\right) = \log_2\left(1 + \frac{\text{SNR}}{\Gamma}\right), \tag{5.78}$$

where

$$\Gamma \triangleq \frac{1}{3}\eta^2. \tag{5.79}$$

The new quantity Γ is usually designated by 'SNR gap' because it represents how far the system is from achieving the channel capacity $C = \log_2(1 + \text{SNR})$, that is, it is a measure of loss with respect to theoretical optimum performance. To obtain a Symbol Error Rate (SER) of $P_s = 2 \times 10^{-7}$ [Cio91], it is necessary that $\eta^2 = 14.5$ dB in (5.73), and therefore $\Gamma = 9.8$ dB. To ensure an adequate performance in the presence of unforseen channel imperfections, an extra margin is required, which we denote by γ_m and designate by system margin. In a system with coding, we also have to subtract the coding gain γ_c. Hence, the SNR gap can be approximated by [Cio91]

$$\Gamma = 9.8 + \gamma_m - \gamma_c \text{ dB}. \tag{5.80}$$

In the of case cross QAM constellations with 2^{2m+1} points, with $m \geq 2$, it is shown in Appendix D that

$$\varepsilon = \frac{31 \times 2^{2m-4} - 1}{6} d^2 \tag{5.81}$$

and

$$P_s \approx 4\left(1 - \frac{3}{2^{m+2}}\right) Q(\eta). \tag{5.82}$$

In this case,

$$\eta = \sqrt{\beta_{\text{cross}}\text{SNR}}, \tag{5.83}$$

with

$$\beta_{\text{cross}} \triangleq \frac{3}{31 \times 2^{2m-4} - 1}. \tag{5.84}$$

Note that $\beta_{\text{cross}} \approx \beta_{\text{square}}$; hence, the preceding analysis for the number of bits and Γ values can also be used with cross QAM constellations.

5.4.2.2 *Multicarrier Analysis*

Since different constellations can be used in different subcarriers, the symbol error rate for the kth subcarrier can be written as

$$P_{s,k} \approx \alpha_k Q(\eta_k), \tag{5.85}$$

with

$$\eta_k \triangleq \frac{d_{\min,k}|H_k|}{\sigma_{N,k}} = \sqrt{\beta_k \text{SNR}_k}, \tag{5.86}$$

where $d_{\min,k}$ denotes the minimum Euclidean distance between points of the constellation used in subcarrier k, $\sigma_{N,k}^2$ represents the variance of the noise on subcarrier k, and parameters α_k and β_k depend on the adopted constellation. SNR_k denotes the signal-to-noise ratio on subcarrier k, which is given by

$$\text{SNR}_k = \frac{E[|S_k|^2]|H_k|^2}{E[|N_k|^2]}. \tag{5.87}$$

Clearly,

$$\text{SNR}_k = \eta_G \text{SNR}^C |H_k|^2, \tag{5.88}$$

where SNR^C denotes the signal-to-noise ratio of the received signal and

$$\eta_G = \frac{\int_0^T w^2(t)dt}{\int_{-\infty}^{+\infty} w^2(t)dt} = \frac{T}{T + T_G}. \tag{5.89}$$

Representing by N_{on} the number of subcarriers turned on and defining $\mathcal{I}$ as the set containing the corresponding indices, the total number of bits transmitted in a multicarrier system is given by

$$b = \sum_{k \in \mathcal{I}} b_k, \tag{5.90}$$

where b_k represents the number of bits transmitted in each subcarrier. For a system using square and cross QAM constellations, it can be shown from (5.72) and (5.81) that b_k can be written as

$$b_k = \log_2 \left(1 + \frac{\text{SNR}_k}{\Gamma}\right). \tag{5.91}$$

Defining a signal-to-noise ratio 'geometric mean' for the overall multicarrier system as

$$\overline{\text{SNR}} \triangleq \left(\prod_{k \in \mathcal{I}} \text{SNR}_k\right)^{1/N_{on}}, \tag{5.92}$$

the total number of bits can be written as

$$b = N_{on} \log_2 \left(1 + \frac{\overline{\text{SNR}}}{\Gamma}\right). \tag{5.93}$$

This expression allows us to calculate the system margin from

$$\gamma_m = 10 \log_{10} \left(\frac{\overline{\text{SNR}}}{2^{b/N_{on}} - 1}\right) - 9.8 + \gamma_c \text{ dB}. \tag{5.94}$$

5.4.2.3 Waterfilling

The energy distribution among active subcarriers is crucial to maximize the systems data rate. The optimal distribution is called waterfilling [Gal68] and is obtained as follows. Defining the coefficients

$$g_k \triangleq \frac{E[|N_k|^2]}{|H_k|^2} \tag{5.95}$$

allows us to write

$$\mathrm{SNR}_k = \frac{\varepsilon_k}{g_k}, \tag{5.96}$$

where $\varepsilon_k = E[|S_k|^2]$ denotes the energy transmitted on subcarrier k. Using (5.90), the problem of assigning bits to subcarriers and distributing the available energy can be written as

$$\max_{\varepsilon_k \in \mathbb{R}_0^+} \sum_{k \in \mathcal{I}} \log_2 \left(1 + \frac{\varepsilon_k}{\Gamma g_k} \right), \tag{5.97}$$

subject to

$$\varepsilon_{\mathrm{tot}} = \sum_{k \in \mathcal{I}} \varepsilon_k, \tag{5.98}$$

where $\varepsilon_{\mathrm{tot}}$ represents the total energy. The solutions obtained for this optimization problem using Lagrange multipliers and Kuhn-Tucker conditions are

$$\varepsilon_k = (\xi - \Gamma g_k)^+, \tag{5.99}$$

where ξ is a constant that ensures that condition (5.98) is satisfied, and $(x)^+ = x$ if $x \geq 0$ and 0 otherwise (i.e., if $\xi - \Gamma g_k < 0$, then $\varepsilon_k = 0$, which means that the subcarrier is turned off). Figure 5.26 graphically illustrates this solution.

5.4.2.4 Leke–Cioffi Algorithm

The first step of the algorithm proposed in [LC97] is to determine which subcarriers are turned on and off. Initially, all subcarriers are considered to be turned on, and the set $\mathcal{I}$ is initialized with all subcarriers. Subcarrier k is turned off if

$$g_k > \frac{\xi}{\Gamma}, \qquad k \in \mathcal{I}, \tag{5.100}$$

where the constant level ξ at each iteration can be obtained from (5.99) simply by adding for all active subcarriers, thus getting

$$\xi = \frac{1}{N_{on}} \left(\varepsilon_{\mathrm{tot}} + \Gamma \sum_{k' \in \mathcal{I}} g_{k'} \right). \tag{5.101}$$

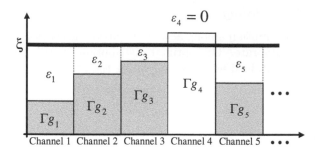

Figure 5.26: Waterfilling energy distribution.

After identifying which subcarriers are turned on/off, the next step is to distribute the total available energy by the active subcarriers. This can be done by simply dividing the available energy equally by the active subcarriers, which is usually called 'flat-energy' distribution. As seen before, the 'waterfilling' distribution is more efficient, since it divides the available energy according to the subchannel capacity. In this case, the energy in subcarrier k is calculated from

$$\varepsilon_k = \xi - \Gamma g_k, \qquad k \in \mathcal{I}. \tag{5.102}$$

Note that set $\mathcal{I}$ has been updated and is now different from the one used in (5.100). The number of bits transmitted in subcarrier k can then be calculated from

$$b_k = \log_2\left(1 + \frac{\varepsilon_k}{\Gamma g_k}\right), \qquad k \in \mathcal{I}. \tag{5.103}$$

The obtained number of bits should be rounded according to the used constellations and the energy recalculated to ensure that the total distributed energy is equal to the total available energy.

5.4.3 Loading with Nonlinear Distortion Effects

As seen in Subsection 5.4.1, employing the statistical characterization of the frequency-domain block to be transmitted, we can calculate an ESNR for each subcarrier, given by (5.71). Defining the coefficients

$$g_k^C \triangleq \frac{|F_k H_k|^2 E[|D_k|^2] + E[|N_k|^2]}{|\alpha F_k H_k|^2} \tag{5.104}$$

this relation can be written as

$$\text{ESNR}_k = \frac{\varepsilon_k}{g_k^C}. \tag{5.105}$$

Using this ratio, we can redefine the loading algorithm presented in the previous subsection and use it to analyze the effects of nonlinear distortion in multicarrier transmission. We can also define an 'equivalent signal-to-noise ratio geometric mean' as

$$\overline{\text{ESNR}} \triangleq \left(\prod_{k \in \mathcal{I}} \text{ESNR}_k \right)^{1/N_{on}} \tag{5.106}$$

and use it in (5.93) and (5.94) to calculate the system margin in the presence of nonlinear distortion effects.

5.4.4   Performance Results

In this subsection, we present a set of results concerning the application of the loading algorithm presented in the previous subsection, taking into consideration nonlinear distortion effects. We consider a multicarrier system, with $N = 256$ subcarriers, $\Gamma = 9.8$ dB and $\gamma_c = 0$. The nonlinear operation is chosen to be an ideal envelope clipping, with clipping level $s_M/\sigma = 2.0$, unless otherwise stated. The multiplying coefficients F_k for $k = 0, 1, ..., N' - 1$ are set to 1 for the N in-band subcarriers and 0 for the out-of-band subcarriers. We assume the use of both square and cross QAM constellations and the transmission over three different frequency-selective channels characterized by the set of coefficients $|H_k^{(1)}|^2$, $|H_k^{(2)}|^2$ and $|H_k^{(3)}|^2$, as depicted in Figure 5.27.

Figures 5.28 and 5.29 show the evolution of $E[|D_k|^2]$ for different values of s_M/σ and SNR^C, respectively, before and after loading has been performed. Figure 5.28 compares self-interference values for several values of the clipping level and the three considered channels, while Figure 5.29 shows the impact of SNR^C on $E[|D_k|^2]$ values. From these figures, it is clear that the channel response influences self-interference levels after loading and in an unpredictable way. This is due to turned off subcarriers and the energy/power assignment. It can be seen from Figure 5.28 that interference levels can be higher or lower for different channels and that the symmetry about the center of the spectrum is lost with fluctuations below 0.5 dB. From Figure 5.29, it is clear that SNR^C values also influence self-interference levels, with differences again depending on the channel response. For example, for channel 1, differences lay in the range]-0.63,0.63[dB and for channel 2 in]-0.35,0.52[dB. Although

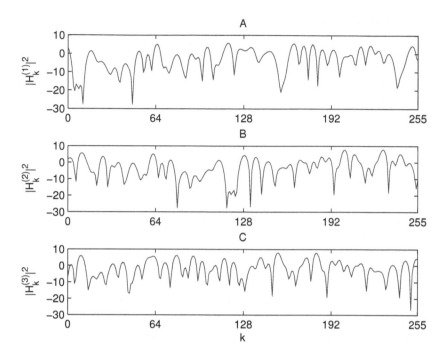

Figure 5.27: Evolution of $|H_k^{(1)}|^2$, $|H_k^{(2)}|^2$ and $|H_k^{(3)}|^2$.

this behaviour is unpredictable, as can be seen with $\mathrm{SNR}^C = 10$ dB, with higher interference levels for channel 1 and lower for channel 2, differences are below 0.5 dB.

Let us now consider the impact of nonlinear distortion effects on the system. We will consider three different scenarios:

I Plain loading;

II Loading taking nonlinear distortion effects into consideration, i.e., using (5.104) instead of (5.95);

III Recalculate self-interference levels using the energy distribution obtained with loading (which changes (2.63)) and repeat loading using this distribution.

When plain loading is performed (scenario I) for channel 1 with $\mathrm{SNR}^C = 15$ dB, we obtain 239 active subcarriers, a total of 1331 bits, and a 1.67 dB system margin. Taking nonlinear distortion effects into consideration affects these values, especially for low values of s_M/σ, as can be

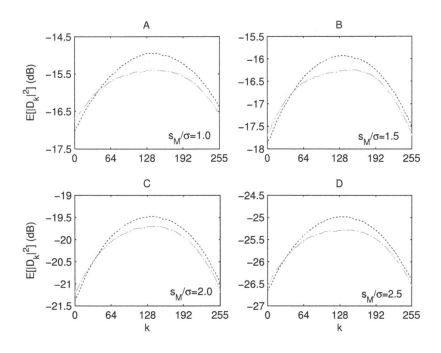

Figure 5.28: Evolution of $E[|D_k|^2]$ **before (solid line) and after loading for channel 1 (dotted line), channel 2 (dash-dotted line) and channel 3 (dashed line), with** $\text{SNR}^C = 15$ **dB and** $s_M/\sigma = 1.0$ **(A),** $s_M/\sigma = 1.5$ **(B),** $s_M/\sigma = 2.0$ **(C) and** $s_M/\sigma = 2.5$ **(D).**

seen in Table 5.1 and Figures 5.30 and 5.31 for channel 1. The real margin of the system in the presence of nonlinear distortion effects, obtained using (5.106) in (5.93) and (5.94), is shown in column NL. Columns labeled II show the values obtained when loading is done taking nonlinear distortion effects into consideration (scenario II). Since loading changes $E[|S_k|^2]$, and, consequently, $E[|D_k|^2]$ (see Figures 5.28 and 5.29), loading can be repeated to obtain more accurate results. This is what we called scenario III, and the obtained values are shown in columns labeled III. The number of active subcarriers is not shown, since it does not change (being 238 in all the considered scenarios).

The impact of s_M/σ and SNR^C on the total number of bits and active subcarriers is shown in Figure 5.30. The impact of s_M/σ and SNR^C on the total number of bits, active subcarriers and margin of the system is shown in Figures 5.30 and 5.31. It can be seen that higher SNR^C values need higher clipping levels to reach the number of bits obtained using

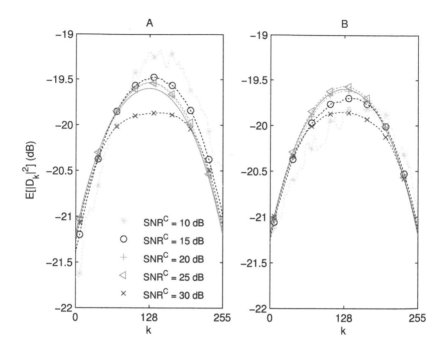

Figure 5.29: Evolution of $E[|D_k|^2]$ before (solid line) and after (dotted line) loading for channel 1 (A) and channel 2 (B) and different values of SNR^C.

plain loading. It can also be seen that reloading does not significantly change the obtained values. Figure 5.31.A shows the impact of nonlinear distortion effects on the system margin. As said before, the margin in the presence of nonlinear effects is obtained using $\overline{\text{ESNR}}$ in (5.94) instead of $\overline{\text{SNR}}$. It can be seen that the values obtained not taking nonlinear distortion effects into consideration can be very optimistic, especially for low clipping values. Figure 5.31.B compares margins values obtained with scenarios I, II and III, and Figure 5.32 illustrates the differences obtained in the number of bits per subcarrier and energy assignment to the active subcarriers with and without nonlinear distortion effects for the first 64 subcarriers of channel 1.

Table 5.1: System Margins Obtained Using (5.106) **and for Scenarios II and III and Total Number of Bits for Scenarios II and III**

s_M/σ	NL	Margin		Number of Bits	
		Scenario II	Scenario III	Scenario II	Scenario III
1.0	-3.19	1.43	1.45	955	949
1.5	-1.33	1.51	1.54	1101	1095
2.0	0.31	1.55	1.58	1223	1219
2.5	1.26	1.44	1.43	1307	1307

5.5 Impact of Nonlinear Signal Processing on OFDMA Signals

This section is dedicated to the analytical evaluation of nonlinear distortion effects on OFDMA signals. We consider nonlinear distortion effects that are inherent to nonlinear signal processing techniques for reducing the PMEPR of the transmitted signals. For this purpose, we take advantage of the Gaussian-like nature of OFDMA signals to assess nonlinear distortion effects in the performance of OFDMA schemes. Our results allow an analytical spectral characterization of the transmitted signals, as well as the computation of the nonlinear interference levels on the received signals. This allows an efficient approach for studying aspects such as the type of nonlinear device, the impact of the system load, or the carrier assignment schemes.

The basic results from this section were published in [AD08b, AD08a, AD10a].

5.5.1 Nonlinear Effects on OFDMA Signals

We consider an OFDMA system with P users and N_p subcarriers assigned to the pth user, as described in Subsection 2.3.2. The total number of subcarriers is N', the number of in-band subcarriers is N, and we have $N' - N$ subcarriers that are always idle. Figure 2.16 illustrates a system with p users in the uplink case.

We consider nonlinear signal processing schemes that operate on a sampled version of the OFDMA signal. Moreover, these nonlinear operations are usually intentional (e.g., clipping and filtering techniques for reducing the PMEPR of the transmitted signals [DG04]). We will focus on the downlink transmission, but a similar analysis can be used for the uplink case. We just have to replace the samples s_n by $s_n^{(p)}$, thus getting the characterization for the P uplink transmitters.

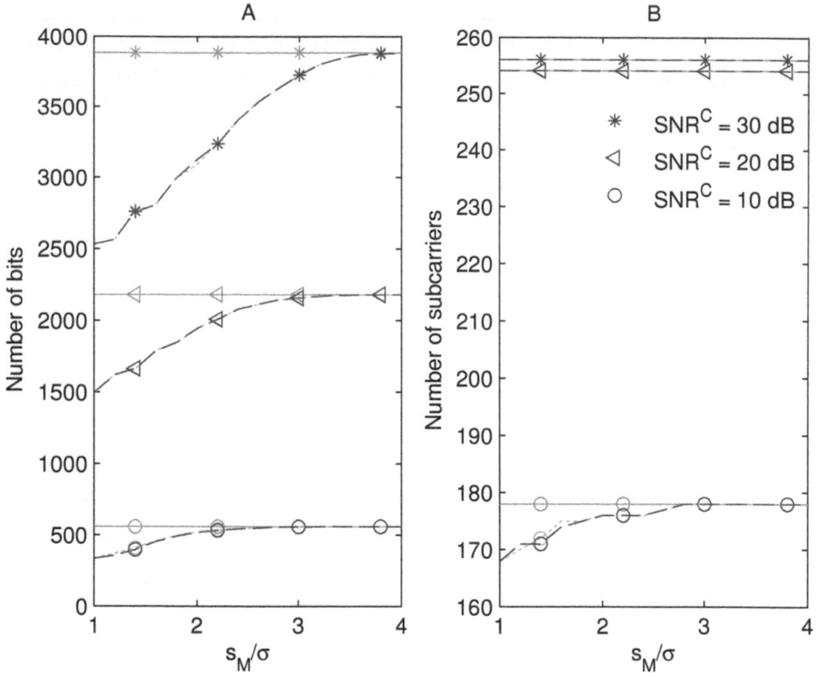

Figure 5.30: Evolution of the total number of bits (A) and the number of active subcarriers (B) for channel 1 and scenario I (solid line), scenario II (dotted line) and scenario III (dashed line).

5.5.2 Statistical Characterization of the Transmitted Signals

The basic downlink transmitter structure considered in this section is similar to the one depicted in Figure 5.25. The nonlinear device is modeled as a polar memoryless nonlinearity operating on an oversampled version of the OFDMA signal, leading to (see (3.22))

$$s_n^C = g(s_n) = f(|s_n|)\, e^{j\,\arg(s_n)}, \tag{5.107}$$

with $f(R) = A(R)\, e^{j\Theta(R)}$. For the uplink transmission, the set of coefficients $\{F_k; k = 0, 1, \dots, N'-1\}$ can be used to eliminate the out-of-band

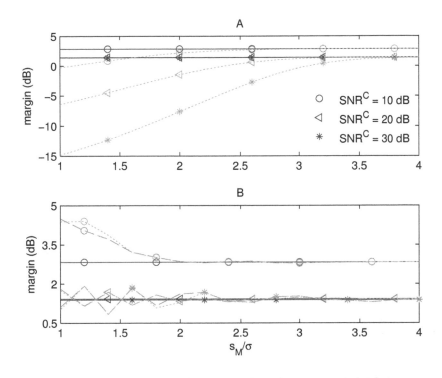

Figure 5.31: Evolution of margin values for channel 1 and scenario I using (5.92) (solid line) and (5.106) (dotted line) **(A)** and for scenario I (solid line), scenario II (dotted line) and scenario III (dashed line) **(B)**.

radiation[4] as well as the interference levels between users[5], both introduced by the nonlinear device. For the downlink transmission, the set of coefficients $\{F_k; k = 0, 1, \ldots, N' - 1\}$ can only be used to eliminate the out-of-band radiation, not the nonlinear interference between users.

As in previous sections, we will take advantage of the Gaussian nature of OFDMA signals with a large number of subcarriers for the analytical characterization of the transmitted signals. Since we are considering a polar memoryless nonlinearity, the analysis made in Section 4.2 will be

---

[4]Throughout this section, the 'in-band' subcarriers are the N subcarriers that can be assigned to users, and the 'out-of-band' subcarriers are the $N' - N$ idle subcarriers associated to the oversampling.

[5]Due to the OFDMA nature of the signals, of the N in-band subcarriers of the overall signal, only N_p are in fact in-band subcarriers of the pth user. The remaining $N - N_p$ subcarriers are outside its in-band region.

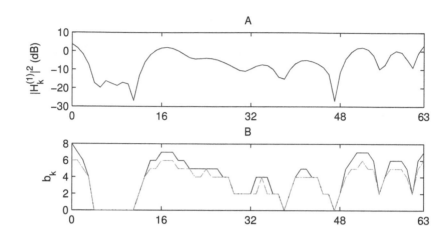

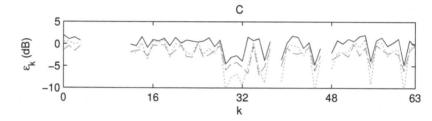

Figure 5.32: Evolution of $|H_k^{(1)}|^2$ (**A**), number of bits per subcarrier (**B**) and energy distribution (**C**), for scenario I (solid line), scenario II (dotted line) and scenario III (dashed line) for the first 64 subcarriers of channel 1 with $\text{SNR}^C = 15$ dB and $s_M/\sigma = 1.5$.

applied. The output of the memoryless nonlinear device can be written as

$$s_n^C = \alpha s_n + d_n, \tag{5.108}$$

where $E[s_n d_{n'}^*] = 0$ and α is given by (4.46), with $R = |s_n|$.

The autocorrelation of the output samples can be expressed as a function of the autocorrelation of the input samples, as in (4.48). The PSD of the transmitted samples is simply obtained from (4.58).

5.5.3 Signal-to-Interference Ratio Levels on the Received Signals

5.5.3.1 Multiple Users

Let us consider the transmission of the OFDMA downlink signal over a frequency-selective channel. It is clear that the symbol received by the pth user on the kth subcarrier is

$$Y_k^{(p)} = \alpha \xi_p F_k H_k^{(p)} \tilde{S}_k^{(p)} + F_k H_k^{(p)} D_k + N_k^{(p)}, \qquad p = 1, \dots, P, \quad (5.109)$$

with $H_k^{(p)}$ and $N_k^{(p)}$ denoting the channel frequency response and channel noise on channel p, respectively. We can calculate a SIR on each subcarrier, given by

$$\mathrm{SIR}_k^{(p)} = \frac{|\alpha|^2 \xi_p^2 E[|\tilde{S}_k^{(p)}|^2]}{E[|D_k|^2]}, \qquad p = 1, \dots, P, \quad (5.110)$$

and also an ESNR on each subcarrier, given by

$$\mathrm{ESNR}_k^{(p)} = \frac{|\alpha|^2 \xi_p^2 |F_k|^2 |H_k^{(p)}|^2 E[|\tilde{S}_k^{(p)}|^2]}{|F_k|^2 |H_k^{(p)}|^2 E[|D_k|^2] + E[|N_k^{(p)}|^2]}, \qquad p = 1, \dots, P. \quad (5.111)$$

When transmitting an OFDMA uplink signal over frequency-selective channels, the received symbol on the kth subcarrier is

$$Y_k^{(p)} = \alpha_p \xi_p F_k^{(p)} H_k^{(p)} \tilde{S}_k^{(p)} + \sum_{p'=1}^{P} F_k^{(p')} H_k^{(p')} D_k^{(p')} + N_k. \quad (5.112)$$

In this case, $\mathrm{SIR}_k^{(p)}$ is given by

$$\mathrm{SIR}_k^{(p)} = \frac{|\alpha_p|^2 \xi_p^2 |F_k^{(p)}|^2 |H_k^{(p)}|^2 E[|\tilde{S}_k^{(p)}|^2]}{I_k^{\mathrm{eq}}}, \quad (5.113)$$

with

$$I_k^{\mathrm{eq}} = \sum_{p'=1}^{P} |F_k^{(p')}|^2 |H_k^{(p')}|^2 E[|D_k^{(p')}|^2], \quad (5.114)$$

and $\mathrm{ESNR}_k^{(p)}$ is given by

$$\mathrm{ESNR}_k^{(p)} = \frac{|\alpha_p|^2 \xi_p^2 |F_k^{(p)}|^2 |H_k^{(p)}|^2 E[|\tilde{S}_k^{(p)}|^2]}{I_k^{\mathrm{eq}} + E[|N_k|^2]}, \qquad p = 1, \dots, P. \quad (5.115)$$

5.5.3.2 Single User

When we have just one active user, e.g., the pth user, on the downlink case, we have

$$Y_k^{(p)} = \alpha \xi_p F_k H_k^{(p)} \tilde{S}_k^{(p)} + F_k H_k^{(p)} D_k + N_k, \tag{5.116}$$

which gives

$$\mathrm{SIR}_k^{(p)} = \frac{|\alpha|^2 \xi_p^2 E[|\tilde{S}_k^{(p)}|^2]}{E[|D_k|^2]} \tag{5.117}$$

and

$$\mathrm{ESNR}_k^{(p)} = \frac{|\alpha|^2 \xi_p^2 |F_k|^2 |H_k^{(p)}|^2 E[|\tilde{S}_k^{(p)}|^2]}{|F_k|^2 |H_k^{(p)}|^2 E[|D_k|^2] + E[|N_k|^2]}. \tag{5.118}$$

For the uplink, we have a similar situation, with

$$Y_k^{(p)} = \alpha_p \xi_p F_k^{(p)} H_k^{(p)} \tilde{S}_k^{(p)} + F_k^{(p)} H_k^{(p)} D_k^{(p)} + N_k, \tag{5.119}$$

for $k \in \Psi_p$ and zero elsewhere. This means that

$$\mathrm{SIR}_k^{(p)} = \frac{|\alpha_p|^2 \xi_p^2 E[|\tilde{S}_k^{(p)}|^2]}{E[|D_k^{(p)}|^2]} \tag{5.120}$$

and

$$\mathrm{ESNR}_k^{(p)} = \frac{|\alpha_p|^2 \xi_p^2 |F_k^{(p)}|^2 |H_k^{(p)}|^2 E[|\tilde{S}_k^{(p)}|^2]}{|F_k^{(p)}|^2 |H_k^{(p)}|^2 E[|D_k^{(p)}|^2] + E[|N_k|^2]}. \tag{5.121}$$

5.5.4 Performance Results

In this section, we present a set of results concerning the performance evaluation of the proposed nonlinear signal processing schemes that operate on a sampled version of the OFDMA signal. The considered OFDMA system has $N = 256$ subcarriers and a Quadrature Phase-Shift Keying (QPSK) constellation, with a Gray mapping rule, on each subcarrier. The system has $P = 4$ users, not necessarily active, with an equal number of subcarriers assigned to each user, i.e., $N_p = 64$, $p = 1, 2, 3, 4$. We consider an ideal envelope clipping, with clipping level $s_M/\sigma = 2.0$, unless otherwise stated. With this clipping level, the 'PMEPR exceeded only 0.1% of the time' (which is almost independent of the number of used subcarriers [DG04]), is about 5.7 dB when the F_k coefficients are used to remove the nonlinear distortion in the unused subcarriers; for $F_k = 1$ (no post-clipping filtering), the PMEPR is about 3.5 dB, and without clipping, the PMEPR is about 8.4 dB. Unless otherwise stated, the oversampling factor is $M_{\mathrm{Tx}} = 2$, and the multiplying coefficients $\{F_k; k = 0, 1, ..., N' - 1\}$ and $\{F_k^{(p)}; k = 0, 1, ..., N' - 1\}$ are set to 1 for the used subcarriers and 0 for the unused and out-of-band subcarriers.

5.5.4.1 Downlink Transmission

As seen in Subsection 2.3.2, in the case of downlink transmission, we consider two different ways of assigning subcarriers to a given user: Regular Grid (RG) and Block Assignment (BA), as shown in Figure 2.15.

In Figure 5.33, we compare values of $E[|D_k|^2]$, measured in dB, obtained using the analytical approach presented in this chapter, with the simulated ones. We consider downlink transmission and different cases concerning the way the power is assigned to the users. The figure clearly shows a good match between theory and simulation, even when we have only one user (64 active subcarriers). Naturally, the accuracy of our analytical approach decreases when the number of active subcarriers is very low.

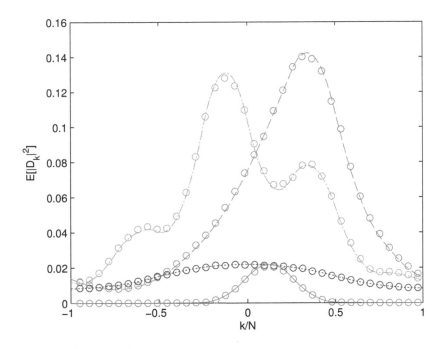

Figure 5.33: Comparison of simulated (o) and analytical values of $E[|D_k|^2]$ **for downlink transmission with four equal power 0 dB users (dotted line), just one 0 dB user active (solid line) and BA with** $\{\xi_1, \xi_2, \xi_3, \xi_4\} = \{-10, 20, 0, 10\}$ **dB (dash-dotted line) or** $\{\xi_1, \xi_2, \xi_3, \xi_4\} = \{-10, 0, 10, 20\}$ **dB (dashed line).**

A comparison of results obtained for self-interference values with and without the use of oversampling can be found in Figure 5.34 for different number of users (i.e, different systems loads), with equal power users and block assignment. It can be seen that we can improve the SIR levels for a fully loaded system (all users active) and two users active by employing an oversampling factor of 2 (negligible improvements are observed with an oversampling factor of 4). If we have just one user active, oversampling is not necessary (i.e., the SIR levels are almost independent of the oversampling factor when we have a single user). Total power of the nonlinear distortion component is independent of the oversampling factor (it is only a function of the normalized clipping level), but part of it is outside the used part of the spectrum when we have oversampling (see also [DG04]); the same applies when the system is not fully loaded. The increase in the values of $E[|D_k|^2]$ at the edge of the spectrum (N' subcarriers) observed in (B) and (C) for $M_{\text{Tx}} = 1$ and $M_{\text{Tx}} = 2$ is due to the aliasing effect. For just one user, the values of $E[|D_k|^2]$ are similar to the ones in the uplink case. Figure 5.35 shows the impact of the normalized clipping level on $E[\text{SIR}_k^{(p)}]$, with equal power users for different system loads and oversampling factors. It can be seen that the improvement obtained by using an oversampling factor of 2 is independent of the clipping level. We can notice again that when there is just one user active, the SIR levels are almost independent of oversampling.

Figure 5.36 compares the variation of $|\alpha|^2 \xi_p^2 E[|\tilde{S}_k^{(p)}|^2]$ and $E[|D_k|^2]$ for block assignment and the two different user assignment of the power control coefficients $\{-10, 0, 10, 20\}$ dB used before. From the figure, it is clear that increasing power of the users within the band leads to significantly lower levels of self-interference and that low-power users can suffer high levels of interference when they are between users with higher assigned powers. Figures 5.37 to 5.39 concern the comparison of two subcarrier assigning schemes: RG and BA. Figure 5.37 compares $E[\text{SIR}_k^{(p)}]$ for the two schemes and different power users. The SIR levels shown concern the -10 dB-power and the 0 dB-power users. Equal power users and the two different user assignment of the power control coefficients $\{-10, 0, 10, 20\}$ dB used before are considered. Figures 5.38 and 5.39 show the impact of subcarrier assignment on $\text{SIR}_k^{(p)}$ and $E[|D_k|^2]$, respectively. Again, we consider two different user assignment of the same power control coefficients. Clearly, the RG scheme allows better SIR levels for users with lower power, although slightly worse for users with higher power. As in Figure 5.36, it can be seen that the performance of low-power users can be severely affected when they are between users with higher assigned powers.

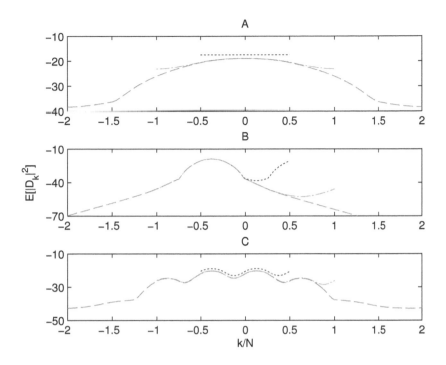

Figure 5.34: Comparison of $E[|D_k|^2]$ **for downlink transmission with equal power users and BA with all users active (A), user 1 active (B) and users 1 and 3 active (C), for** $M_{\mathrm{Tx}} = 1$ **(dotted line),** $M_{\mathrm{Tx}} = 2$ **(dash-dotted line) and** $M_{\mathrm{Tx}} = 4$ **(dashed line).**

5.5.4.2   Uplink Transmission

In the uplink transmission, post-clipping filtering is generally used to eliminate the out-of-band radiation. Figure 5.40 compares the variation of $\mathrm{SIR}_k^{(p)}$ for different assignment of the power control coefficients $\{-10, 0, 10, 20\}$ dB in case no post-clipping filtering is used. Again, these coefficients are assigned in two different ways: no particular order or increasing order. Clearly, we have a behaviour similar to the one observed for the downlink transmission, namely worse SIR levels for low-power users, especially when they use sets of subcarriers that are adjacent to high-power subcarriers (i.e., when we have low-power users close to high-power users). Naturally, if we have a post-clipping filtering able to eliminate the nonlinear distortion for the subcarriers that are not used to a given user, then the SIR levels of a given user are only a function

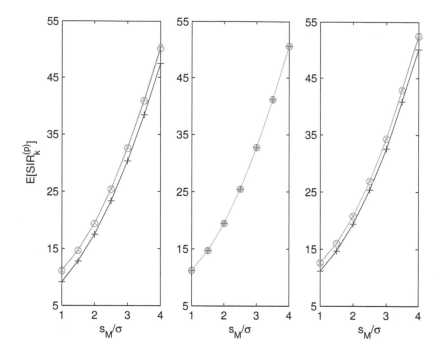

Figure 5.35: Variation of $E[\text{SIR}_k^{(p)}]$ for downlink transmission and BA with equal power users with all users active (A), user 1 active (B) and users 1 and 3 active (C), for $M_{\text{Tx}} = 1$ (+), $M_{\text{Tx}} = 2$ (∘) and $M_{\text{Tx}} = 4$ (×).

of the clipping for that user, even when we have low-power users adjacent to high-power users. Therefore, in contrast to the downlink case, the post-clipping filtering can increase the system's robustness significantly. Let us now consider the transmission over a frequency-selective channel. Figure 5.41 shows $\text{ESNR}_k^{(p)}$ for the 0 dB-power user with and without post-clipping filtering for different assignment of power control coefficients. The results for 4 equal power users (each with 0 dB) are also included for the sake of comparison. From this figure, it is clear that the ESNR values are much better when we have post-clipping filtering, especially for the subcarriers that have deep fading. This is aggravated for low-power users, especially when they are adjacent to high-power users.

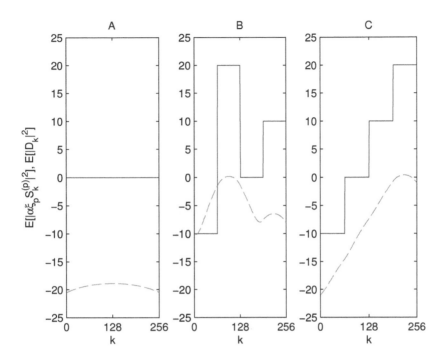

Figure 5.36: Variation of $|\alpha|^2 \xi_p^2 E[|\tilde{S}_k^{(p)}|^2]$ **(solid line) and** $E[|D_k|^2]$ **(dashed line) for downlink transmission and BA with equal 0 dB power users (A),** $\{\xi_1, \xi_2, \xi_3, \xi_4\} = \{-10, 20, 0, 10\}$ **dB (B) and** $\{\xi_1, \xi_2, \xi_3, \xi_4\} = \{-10, 0, 10, 20\}$ **dB (C).**

5.6 Discussion

In Section 5.1, general clipping and filtering techniques for reducing the envelope fluctuations of multicarrier signals were described. These techniques involve a nonlinear clipping operation, followed by a filtering procedure. Results show that envelope clipping (polar clipping) has better power efficiency than real/imaginary clipping (Cartesian clipping). Simulation results show that, whenever $N \geq 64$, the analytical approach for computing the PSD of the transmitted signals is very accurate, regardless the type of clipping, the clipping level, the oversampling factor and the constellation size.

Sections 5.2 and 5.3 are dedicated to the analysis of the impact of two different kinds of quantization effects: the numerical accuracy inherent to the DFT/IDFT on multicarrier signals, and the quantization

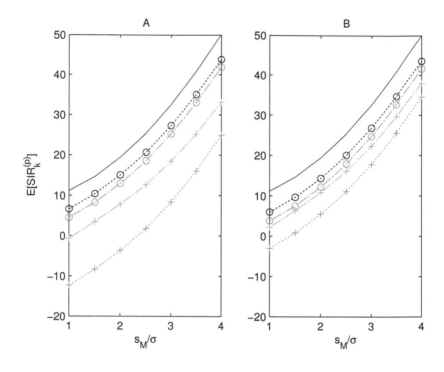

Figure 5.37: Variation of $E[\mathrm{SIR}_k^{(p)}]$ for downlink transmission with equal power 0 dB users (solid line) and RG ($\circ$) or BA ($+$), with $\{\xi_1, \xi_2, \xi_3, \xi_4\} = \{-10, 20, 0, 10\}$ dB (dotted line) or $\{\xi_1, \xi_2, \xi_3, \xi_4\} = \{-10, 0, 10, 20\}$ dB (dash-dotted line), for the users with $\xi_p = -10$ dB (A) and $\xi_p = 0$ dB (B).

requirements within the ADC used in software radio architectures. An analytical approach for analyzing this impact was presented and used for the performance evaluation of given quantization characteristics, in a simple and computationally efficient way. In case of quantization effects on multicarrier signals, our results indicate that the quantization requirements are higher at the receiver than at the transmitter, especially for severely frequency-selective channels. It is also shown that the selection of the oversampling factor and the ΔN factor can have a significant impact on the SIR levels. As for software radio signals, our results show that the oversampling factor and the reference frequency used for defining the complex envelopes can have a significant effect on the ADC performance, allowing improvements on the SIR levels of several decibels.

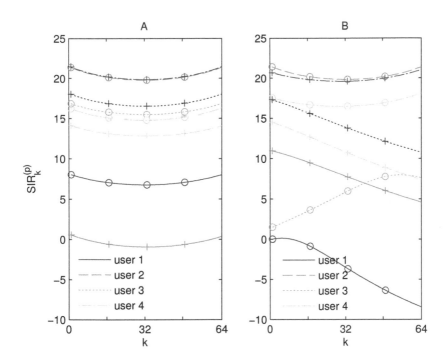

Figure 5.38: Comparison of $\mathbf{SIR}_k^{(p)}$ for downlink transmission with RG (A) and BA (B) with $\{\xi_1, \xi_2, \xi_3, \xi_4\} = \{-10, 20, 0, 10\}$ dB ($\circ$) or $\{\xi_1, \xi_2, \xi_3, \xi_4\} = \{-10, 0, 10, 20\}$ dB ($+$).

It should be noted that our statistical approach could be employed with other Cartesian memoryless nonlinearities, such as Cartesian clipping for reducing the envelope fluctuations of the transmitted signals [DG01]. Therefore, there are also gains with the oversampling factor and the ΔN factor on those cases.

In Section 5.4, we studied the impact of nonlinear distortion effects on adaptive multicarrier systems. We included an analytical statistical characterization of the transmitted signals, which was used to evaluate loading algorithms in the presence of strong nonlinear distortion effects and to redefine loading algorithms taking into account nonlinear distortion issues. A set of performance results was presented, showing that the margin of multicarrier systems can be improved by taking nonlinear distortion effects into consideration when applying loading algorithms.

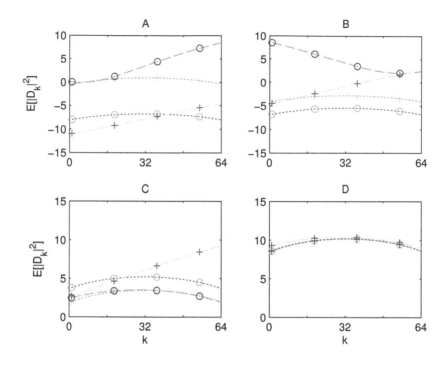

Figure 5.39: Comparison of $E[|D_k|^2]$ **for downlink transmission with RG (dotted line) and BA (dashed line), with** $\{\xi_1, \xi_2, \xi_3, \xi_4\} = \{-10, 20, 0, 10\}$ **dB** (○) **or** $\{\xi_1, \xi_2, \xi_3, \xi_4\} = \{-10, 0, 10, 20\}$ **dB** (+)**, for the user with** $\xi_p = -10$ **dB** (A)**,** $\xi_p = 0$ **dB** (B)**,** $\xi_p = 10$ **dB** (C) **and** $\xi_p = 20$ **dB** (D)**.**

More accurate bit and energy distributions per subcarrier were also obtained.

An analytical tool to evaluate nonlinear distortion effects on systems employing OFDMA signals is presented in Section 5.5. Our results allow an analytical spectral characterization of the transmitted signals, as well as the computation of the nonlinear interference levels on the received signals. A set of performance results was presented, showing that the power allocated to each user has a key impact on the nonlinear distortion effects. Users with smaller allocated power face stronger interference levels, and increasing assignment of power control coefficients can lead to lower interference levels. It is also shown that nonlinear distortion levels are significantly different when just a small fraction of the subcarriers is used (i.e., when the system load is small).

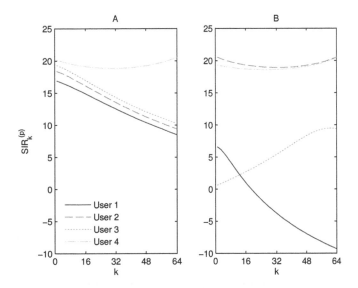

Figure 5.40: Comparison of $\mathrm{SIR}_k^{(p)}$ for uplink transmission without post-clipping filtering with $\{\xi_1, \xi_2, \xi_3, \xi_4\} = \{-10, 0, 10, 20\}$ dB (A) and $\{\xi_1, \xi_2, \xi_3, \xi_4\} = \{-10, 20, 0, 10\}$ dB (B).

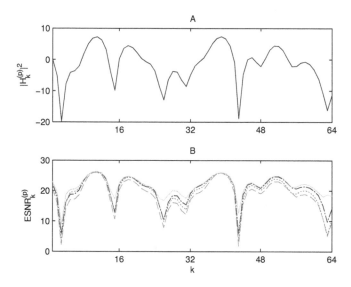

Figure 5.41: Evolution of $|H_k^{(p)}|^2$ (A) and comparison of ESNR$_k^{(p)}$ (B) for uplink transmission with post-clipping filtering (solid line) and without post-clipping filtering and equal power 0 dB users (dash-dotted line), $\{\xi_1, \xi_2, \xi_3, \xi_4\} = \{-10, 20, 0, 10\}$ dB (dashed line) or $\{\xi_1, \xi_2, \xi_3, \xi_4\} = \{-10, 0, 10, 20\}$ dB (dotted line), for the user with $\xi_p = 0$ dB.

Appendix A

Output Autocorrelation for Polar Memoryless Nonlinearities with Gaussian Inputs

In this appendix, we calculate the autocorrelation of the signal at the output of a polar memoryless nonlinearities when the input signal is Gaussian.

Let us consider a bandpass Gaussian signal with real part $x_{BP}(t)$ given by (3.8), and let $x(t)$ denote its complex envelope, i.e.,

$$x(t) = x_I(t) + jx_Q(t) = R(t)\,e^{j\varphi(t)}, \qquad (A.1)$$

with $R = R(t) = |x(t)|$ and $\varphi = \varphi(t) = \arg(x(t))$, which we assume is a complex stationary Gaussian process with equally distributed, zero mean real and imaginary components.

Its autocorrelation function is

$$\begin{aligned} R_x(\tau) &\triangleq E[x(t)x^*(t-\tau)] \\ &= R_{II}(\tau) + R_{QQ}(\tau) + j(R_{QI}(\tau) - R_{IQ}(\tau)), \qquad (A.2) \end{aligned}$$

with

$$R_{II}(\tau) \triangleq E[x_I(t)x_I(t - \tau)] \tag{A.3a}$$

$$R_{QQ}(\tau) \triangleq E[x_Q(t)x_Q(t - \tau)] = R_{II}(\tau) \tag{A.3b}$$

$$R_{QI}(\tau) \triangleq E[x_Q(t)x_I(t - \tau)] \tag{A.3c}$$

$$R_{IQ}(\tau) \triangleq E[x_I(t)x_Q(t - \tau)] = -R_{QI}(\tau). \tag{A.3d}$$

This means that

$$R_x(\tau) = 2(R_{II}(\tau) + j\,R_{QI}(\tau)) \tag{A.4}$$

and $R_x(0) = E[x^2(t)] = 2R_{II}(0) = 2\sigma^2$, since $R_{II}(0) = R_{QQ}(0) = \sigma^2$ and $R_{IQ}(0) = R_{QI}(0) = 0$.

As seen in Section 3.1, if this signal is submitted to a polar memoryless nonlinearity characterized by $f(R) = A(R)\,e^{j\Theta(R)}$, then the complex envelope of the signal at the output of the nonlinearity can be written as

$$y(t) = g(x(t)) = g(R(t)\,e^{j\varphi(t)}) = f(R(t))\,e^{j\varphi(t)}. \tag{A.5}$$

The autocorrelation function of the signal at the output of the nonlinearity is

$$\begin{aligned}
R_y(\tau) &= E[y(t)y^*(t - \tau)] = E[g(x_1)g^*(x_2)] \\
&= E[g(x_{1I} + jx_{1Q})g^*(x_{2I} + jx_{2Q})] \\
&= \int_{-\infty}^{+\infty}\int_{-\infty}^{+\infty}\int_{-\infty}^{+\infty}\int_{-\infty}^{+\infty} g(x_{1I} + jx_{1Q})g^*(x_{2I} + jx_{2Q}) \\
&\quad \cdot p(x_{1I}, x_{1Q}, x_{2I}, x_{2Q})dx_{1I}dx_{1Q}dx_{2I}dx_{2Q}, \tag{A.6}
\end{aligned}$$

where $x_1 = x(t) = x_{1I} + jx_{1Q}$ and $x_2 = x(t - \tau) = x_{2I} + jx_{2Q}$.

Let us consider the vector $w = [x_{1I}\,x_{1Q}\,x_{2I}\,x_{2Q}]^T$ ($[\cdot]^T$ denotes matrix transpose). The covariance matrix of w is

$$W = E[ww^T] = \begin{bmatrix} \sigma^2 & 0 & R_{II}(\tau) & R_{QI}(\tau) \\ 0 & \sigma^2 & -R_{QI}(\tau) & R_{II}(\tau) \\ R_{II}(\tau) & -R_{QI}(\tau) & \sigma^2 & 0 \\ R_{QI}(\tau) & R_{II}(\tau) & 0 & \sigma^2 \end{bmatrix}, \tag{A.7}$$

whose determinant is

$$\begin{aligned}
\det(W) &= \sigma^8 \left(1 - \frac{R_{II}^2(\tau) + R_{QI}^2(\tau)}{\sigma^4}\right)^2 \\
&= \sigma^8 \left(1 - \frac{|R_x(\tau)|^2}{4\sigma^4}\right)^2 = \sigma^8(1 - \rho^2)^2, \tag{A.8}
\end{aligned}$$

where the correlation coefficient ρ corresponds to the normalized auto-correlation of the input signal and is given by

$$\rho \triangleq \rho(\tau) = \frac{E[x_1 x_2]}{E[x_1^2]} = \frac{|R_x(\tau)|}{R_x(0)} = \frac{|R_x(\tau)|}{2\sigma^2}. \tag{A.9}$$

The inverse of W is

$$W^{-1} = \frac{1}{\sqrt{\det(W)}} \begin{bmatrix} \sigma^2 & 0 & -R_{II}(\tau) & -R_{QI}(\tau) \\ 0 & \sigma^2 & R_{QI}(\tau) & -R_{II}(\tau) \\ -R_{II}(\tau) & R_{QI}(\tau) & \sigma^2 & 0 \\ -R_{QI}(\tau) & -R_{II}(\tau) & 0 & \sigma^2 \end{bmatrix}, \tag{A.10}$$

therefore the joint Probability Density Function (PDF) is given by [Goo63, vdB95]

$$p(w) = \frac{1}{(2\pi)^2 \sqrt{\det(W)}} e^{-\frac{1}{2}w^T W^{-1} w} = \frac{1}{4\pi^2 \sigma^4 (1-\rho^2)} e^A, \tag{A.11}$$

with

$$\begin{aligned} A &= -\frac{1}{2\sqrt{\det(W)}} \big(\sigma^2(x_{1I}^2 + x_{2I}^2 + x_{1Q}^2 + x_{2Q}^2) \\ &\quad - 2R_{II}(\tau)(x_{1I}x_{2I} + x_{1Q}x_{2Q}) - 2R_{QI}(\tau)(x_{1I}x_{2Q} - x_{1Q}x_{2I})\big) \\ &= -\frac{\sigma^2(|x_1|^2 + |x_2|^2) - \text{Re}\{R_x(\tau)x_1 x_2^*\}}{2\sigma^4(1-\rho^2)}. \end{aligned} \tag{A.12}$$

The change to polar coordinates $x_1 = R_1 e^{j\varphi_1}$, $x_2 = R_2 e^{j\varphi_2}$, allows us to rewrite (A.12) as [Ric45, Bla68]

$$\begin{aligned} A &= -\frac{\sigma^2(R_1^2 + R_2^2) - |R_x(\tau)|R_1 R_2 \cos(\varphi_1 - \varphi_2 + \arg(R_x(\tau)))}{2\sigma^4(1-\rho^2)} \\ &= -\frac{R_1^2 + R_2^2 - 2\rho R_1 R_2 \cos(\varphi_1 - \varphi_2 + \phi)}{\rho_0}, \end{aligned} \tag{A.13}$$

with $\phi = \arg(R_x(\tau))$ and $\rho_0 = 2\sigma^2(1 - \rho^2)$. Substituting (A.11) and (A.13) in (A.6) and applying the corresponding polar coordinate transformation, we get

$$\begin{aligned} R_y(\tau) &= \frac{1}{2\pi^2 \sigma^2 \rho_0} \int_0^{+\infty} \int_0^{+\infty} \int_0^{2\pi} \int_0^{2\pi} f(R_1) f^*(R_2) R_1 R_2 \, e^{j(\varphi_1 - \varphi_2)} \\ &\quad \cdot e^{-(R_1^2 + R_2^2 - 2\rho R_1 R_2 \cos(\varphi_1 - \varphi_2 + \phi))/\rho_0} \, dR_1 dR_2 d\varphi_1 d\varphi_2. \end{aligned} \tag{A.14}$$

After some lengthy but straightforward manipulations, we obtain

$$
R_y(\tau) = \frac{1}{2\rho_0 \pi^2 \sigma^2} \int_0^{+\infty} \int_0^{+\infty} f(R_1) f^*(R_2) R_1 R_2 \, e^{-(R_1^2 + R_2^2)/\rho_0}
$$

$$
\cdot \left(\int_0^{2\pi} \int_0^{2\pi} e^{j(\varphi_1 - \varphi_2 + \phi - \phi)} \, e^{2\rho R_1 R_2 \cos(\varphi_1 - \varphi_2 + \phi)/\rho_0} d\varphi_1 d\varphi_2 \right) dR_1 dR_2
$$

$$
= \frac{e^{-j\phi}}{2\rho_0 \pi^2 \sigma^2} \int_0^{+\infty} \int_0^{+\infty} f(R_1) f^*(R_2) R_1 R_2 \, e^{-(R_1^2 + R_2^2)/\rho_0}
$$

$$
\cdot \left(\int_0^{2\pi} \int_0^{2\pi} e^{j(\varphi_1 - \varphi_2 + \phi)} \, e^{2\rho R_1 R_2/\rho_0 \cos(\varphi_1 - \varphi_2 + \phi)} d\varphi_1 d\varphi_2 \right) dR_1 dR_2.
$$

$$(A.15)$$

Let $I_n(z)$ denote modified Bessel functions of first kind, which can be written as (see [AS72], page 376)

$$
I_n(z) = \frac{1}{\pi} \int_0^\pi \cos(n\theta) e^{z \cos \theta} d\theta. \tag{A.16}
$$

By assuming $n = 1$, $z = 2\rho R_1 R_2/\rho_0$ and $\theta = \varphi_1 - \varphi_2 + \phi$, we get

$$
I_1\left(\frac{2\rho R_1 R_2}{\rho_0} \right) = \frac{1}{\pi} \int_0^\pi \cos \theta \, e^{2\rho R_1 R_2/\rho_0 \cos \theta} d\theta. \tag{A.17}
$$

Using (A.17) and noting that the integrand is a periodic function, it is easy to find that

$$
\int_0^{2\pi} e^{j\theta} e^{z \cos \theta} d\theta = \int_0^{2\pi} (\cos \theta + j \sin \theta) \, e^{z \cos \theta} d\theta = 2\pi I_1(z), \tag{A.18}
$$

and the double integral in (A.15) can be calculated as

$$
\int_0^{2\pi} \int_0^{2\pi} \cos(\varphi_1 - \varphi_2 + \phi) \, e^{-2\rho R_1 R_2/\rho_0 \cos(\varphi_1 - \varphi_2 + \phi)} d\varphi_1 d\varphi_2
$$

$$
= 4\pi^2 I_1\left(\frac{2\rho R_1 R_2}{\rho_0} \right). \tag{A.19}
$$

Therefore, (A.15) becomes

$$
R_y(\tau) = \frac{2 \, e^{-j\phi}}{\rho_0 \sigma^2} \int_0^{+\infty} \int_0^{+\infty} f(R_1) f^*(R_2) R_1 R_2 \, e^{-(R_1^2 + R_2^2)/\rho_0}
$$

$$
\cdot I_1\left(\frac{2\rho R_1 R_2}{\rho_0} \right) dR_1 dR_2. \tag{A.20}
$$

Using the Laguerre polynomial series expansion expression known as the Hille-Hardy formula (see [PBM86], Equation (5.11.3.7)),

$$\sum_{\gamma=0}^{+\infty} \frac{1}{\gamma+1} \rho^{2\gamma} L_{\gamma}^{(1)}\left(\frac{R_1^2}{2\sigma^2}\right) L_{\gamma}^{(1)}\left(\frac{R_2^2}{2\sigma^2}\right) = \frac{2\sigma^2}{\rho R_1 R_2 (1-\rho^2)} e^{-\rho^2(R_1^2+R_2^2)/\rho_0}$$

$$\cdot I_1\left(\frac{2\rho R_1 R_2}{\rho_0}\right), \quad \text{(A.21)}$$

with $L_m^{(1)}(x)$ denoting a generalized Laguerre polynomial of order m, which is defined as

$$L_m^{(1)}(x) \triangleq \frac{1}{m!} x^{-1} e^x \frac{d^m}{dx^m}\left(e^{-x} x^{m+1}\right), \quad \text{(A.22)}$$

and replacing it in (A.20), we obtain

$$R_y(\tau) = \frac{2}{\rho_0 \sigma^2} \frac{\rho(1-\rho^2)}{2\sigma^2} e^{-j\phi} \int_0^{+\infty} \int_0^{+\infty} f(R_1) f^*(R_2) R_1^2 R_2^2$$

$$\cdot e^{-\rho^2(R_1^2+R_2^2)/\rho_0} e^{\rho^2(R_1^2+R_2^2)/\rho_0} e^{-(R_1^2+R_2^2)/\rho_0} \frac{2\sigma^2}{\rho R_1 R_2 (1-\rho^2)}$$

$$\cdot I_1\left(\frac{2\rho R_1 R_2}{\rho_0}\right) dR_1 dR_2$$

$$= \frac{\rho}{2\sigma^6} e^{-j\phi} \int_0^{+\infty} \int_0^{+\infty} f(R_1) f^*(R_2) R_1^2 R_2^2$$

$$\cdot e^{-(R_1^2+R_2^2)(1-\rho^2)/(2\sigma^2(1-\rho^2))} e^{-\rho^2(R_1^2+R_2^2)/\rho_0} \frac{2\sigma^2}{\rho R_1 R_2 (1-\rho^2)}$$

$$\cdot I_1\left(\frac{2\rho R_1 R_2}{\rho_0}\right) dR_1 dR_2$$

$$= \frac{1}{2\sigma^6} e^{-j\phi} \int_0^{+\infty} \int_0^{+\infty} f(R_1) f^*(R_2) R_1^2 R_2^2 e^{-(R_1^2+R_2^2)/(2\sigma^2)}$$

$$\cdot \left(\sum_{\gamma=0}^{+\infty} \frac{1}{\gamma+1} \rho^{2\gamma+1} L_{\gamma}^{(1)}\left(\frac{R_1^2}{2\sigma^2}\right) L_{\gamma}^{(1)}\left(\frac{R_2^2}{2\sigma^2}\right)\right) dR_1 dR_2. \quad \text{(A.23)}$$

The double integral in (A.23) is separable in two equal integrals with respect to R_1 and R_2, thus obtaining

$$R_y(\tau) = \frac{1}{2\sigma^6} e^{-j\phi} \sum_{\gamma=0}^{+\infty} \frac{1}{\gamma+1} \rho^{2\gamma+1}$$

$$\cdot \left|\int_0^{+\infty} R^2 f(R) e^{-R^2/(2\sigma^2)} L_{\gamma}^{(1)}\left(\frac{R^2}{2\sigma^2}\right) dR\right|^2. \quad \text{(A.24)}$$

This can be written as

$$R_y(\tau) = 2 \sum_{\gamma=0}^{+\infty} P_{2\gamma+1}\, \rho^{2\gamma+1}\, e^{-j\phi}, \qquad (A.25)$$

with coefficients $P_{2\gamma+1}$ given by

$$P_{2\gamma+1} = \frac{1}{4\sigma^6(\gamma+1)} \left| \int_0^{+\infty} R^2 f(R)\, e^{-R^2/(2\sigma^2)} L_\gamma^{(1)} \left(\frac{R^2}{2\sigma^2}\right) dR \right|^2. \qquad (A.26)$$

These coefficients correspond to the total power associated with the Intermodulation Product (IMP) of order $2\gamma + 1$ (see [Shi71, IS73, Ste74, DG97]). Having in mind (A.9), it is easy to find that

$$\rho^{2\gamma+1}\, e^{-j\phi} = \frac{(R_x(\tau))^{\gamma+1}(R_x^*(\tau))^\gamma}{(R_x(0))^{2\gamma+1}}, \qquad (A.27)$$

and (A.25) becomes

$$R_y(\tau) = 2 \sum_{\gamma=0}^{+\infty} P_{2\gamma+1} f_{2\gamma+1}^R(R_x(\tau)), \qquad (A.28)$$

with

$$f_{2\gamma+1}^R(R(\tau)) \triangleq \frac{(R(\tau))^{\gamma+1}(R^*(\tau))^\gamma}{(R(0))^{2\gamma+1}}. \qquad (A.29)$$

Appendix B

Analytical Results for Specific Nonlinearities

In this appendix, we use results from Sections 3.2 and 3.4 to characterize the signal at the output of specific nonlinearities when the input signal is Gaussian.

We study Cartesian memoryless nonlinearities (see Figure 3.8) and polar memoryless nonlinearities (see Figure 3.5), and, in both cases, consider ideal clipping functions and polynomial functions. For these different cases, we develop expressions for the parameter α, associated with the decomposition of the nonlinearly distorted signal into useful and self-interference components, as well as for the total power at the output of the nonlinearity P_{out} and the power associated with each Intermodulation Product (IMP) $P_{2\gamma+1}$. These expressions will be used in several chapters of this work.

B.1 Cartesian Memoryless Nonlinearities

We will first consider a Cartesian memoryless nonlinear device characterized by function $g(x)$ with a bandpass Gaussian signal as input. In Section 3.2, it was shown that the useful component scale factor α is given by

$$\alpha = \frac{E[xg(x)]}{E[x^2]} = \frac{1}{\sqrt{2\pi}\sigma^3} \int_{-\infty}^{+\infty} xg(x) \, e^{-\frac{x^2}{2\sigma^2}} \, dx, \qquad \text{(B.1)}$$

and the average power of the signal at the nonlinearity output is given by

$$P_{\text{out}} = E[g^2(x)] = \frac{1}{\sqrt{2\pi}\sigma} \int_{-\infty}^{+\infty} g^2(x)\, e^{-\frac{x^2}{2\sigma^2}}\, dx. \tag{B.2}$$

The total power associated to the IMP of order $2\gamma + 1$ is given by

$$P_{2\gamma+1} = \frac{1}{2^{2\gamma+1}(2\gamma+1)!} \left(\int_{-\infty}^{+\infty} g(x)p(x)H_{2\gamma+1}\left(\frac{x}{\sqrt{2}\sigma}\right) dx \right)^2, \tag{B.3}$$

which can be rewritten as

$$P_{2\gamma+1} = \frac{2}{\pi\sigma^2 2^{2\gamma+1}(2\gamma+1)!} \left(\int_{0}^{+\infty} g(x)\, e^{-\frac{x^2}{2\sigma^2}} H_{2\gamma+1}\left(\frac{x}{\sqrt{2}\sigma}\right) dx \right)^2$$

$$= \frac{2}{\pi\sigma^2 2^{2\gamma+1}(2\gamma+1)!} |\nu_{2\gamma+1}|^2, \tag{B.4}$$

with

$$\nu_{2\gamma+1} = \int_{0}^{+\infty} g(x)\, H_{2\gamma+1}\left(\frac{x}{\sqrt{2}\sigma}\right) e^{-\frac{x^2}{2\sigma^2}}\, dx$$

$$= \sigma \int_{0}^{+\infty} g(\sigma x)\, H_{2\gamma+1}\left(\frac{x}{\sqrt{2}}\right) e^{-\frac{x^2}{2}}\, dx. \tag{B.5}$$

For odd values of n, Hermite polynomials can be written explicitly as [AS72]

$$H_n(x) = n! \sum_{m=0}^{(n-1)/2} \frac{(-1)^{(n-1)/2-m}}{(2m+1)!((n-1)/2-m)!}(2x)^{2m+1}. \tag{B.6}$$

Hence, we can write

$$H_{2\gamma+1}\left(\frac{x}{\sqrt{2}}\right) = \sum_{m=0}^{\gamma} h_{\gamma,m}\, x^{2m+1}, \tag{B.7}$$

with

$$h_{\gamma,m} = \frac{(2\gamma+1)!}{(2m+1)!(\gamma-m)!}(-1)^{\gamma-m}(\sqrt{2})^{2m+1}, \tag{B.8}$$

which can be used for calculating $\nu_{2\gamma+1}$.

Next, we present some specific expressions for α, P_{out} e $\nu_{2\gamma+1}$ for the cases of an ideal Cartesian clipping and when $g(x)$ can be written as a polynomial.

Cartesian Ideal Clipping

Let us consider a Cartesian ideal clipping, as shown in Figure 3.8, with $g(x)$ given by (3.3). In this case, α, P_{out} and $\nu_{2\gamma+1}$ can be obtained as follows. Replacing (3.3) in (B.1), we get

$$\alpha = \frac{2}{\sqrt{2\pi}\sigma^3} \int_0^{s_M} x^2 e^{-\frac{x^2}{2\sigma^2}} dx + \frac{2s_M}{\sqrt{2\pi}\sigma^3} \int_{s_M}^{+\infty} x\, e^{-\frac{x^2}{2\sigma^2}} dx$$

$$= \frac{2}{\sqrt{2\pi}} \int_0^{s_M/\sigma} x^2 e^{-\frac{x^2}{2}} dx + \frac{2s_M}{\sqrt{2\pi}\sigma} \int_{s_M/\sigma}^{+\infty} x\, e^{-\frac{x^2}{2}} dx. \qquad \text{(B.9)}$$

Let

$$I_n(x) \triangleq \int x^n\, e^{-\frac{x^2}{2}}\, dx. \qquad \text{(B.10)}$$

It can easily be shown that (see [GR07])

$$I_0(x) = \int e^{-\frac{x^2}{2}} dx = \frac{\sqrt{2\pi}}{2}(1 - 2Q(x)), \qquad \text{(B.11)}$$

where $Q(x)$ is the well-known error function given by (5.34), and

$$I_1(x) = \int x\, e^{-\frac{x^2}{2}} dx = -e^{-\frac{x^2}{2}}. \qquad \text{(B.12)}$$

Using induction and integration by parts, it can also be shown that

$$I_{2m+1}(x) = \int x^{2m+1} e^{-\frac{x^2}{2}} dx = -2^m\, m!\, e^{-\frac{x^2}{2}} \sum_{l=0}^{m} \frac{1}{l!}\left(\frac{x^2}{2}\right)^l, \qquad \text{(B.13)}$$

for $m \geq 0$, and

$$I_{2m}(x) = \int x^{2m} e^{-\frac{x^2}{2}} dx$$

$$= -(2m-1)!!\, e^{-\frac{x^2}{2}} \sum_{l=0}^{m-1} \frac{x^{2m-1-2l}}{(2m-1-2l)!!} + (2m-1)!!I_0(x) \qquad \text{(B.14)}$$

for $m \geq 1$, where $(2m+1)!!$ denotes 'double factorial for odd numbers' and is equal to

$$(2m+1)!! \triangleq \prod_{k=0}^{m}(2k+1). \qquad \text{(B.15)}$$

From (B.14), we can write

$$I_2(x) = -x\, e^{-\frac{x^2}{2}} + \frac{\sqrt{2\pi}}{2}(1 - 2Q(x)), \qquad \text{(B.16)}$$

hence, (B.9) becomes

$$
\alpha = \frac{2}{\sqrt{2\pi}} \left(I_2 \left(\frac{s_M}{\sigma} \right) - I_2(0) \right) + \frac{2s_M}{\sqrt{2\pi}\sigma} \left(\lim_{x \to +\infty} I_1(x) - I_1 \left(\frac{s_M}{\sigma} \right) \right)
$$
$$
= 1 - 2Q \left(\frac{s_M}{\sigma} \right). \tag{B.17}
$$

By replacing (3.3) in (B.2), the power of the signal at the nonlinearity output is expressed as

$$
\begin{aligned}
P_{\text{out}} &= \frac{2}{\sqrt{2\pi\sigma^2}} \int_0^{s_M} x^2 \, e^{-\frac{x^2}{2\sigma^2}} \, dx + \frac{2}{\sqrt{2\pi\sigma^2}} \int_{s_M}^{+\infty} s_M^2 e^{-\frac{x^2}{2\sigma^2}} \, dx \\
&= \frac{2\sigma^2}{\sqrt{2\pi}} \int_0^{s_M/\sigma} x^2 \, e^{-\frac{x^2}{2}} \, dx + \frac{2s_M^2}{\sqrt{2\pi}} \int_{s_M/\sigma}^{+\infty} e^{-\frac{x^2}{2}} \, dx \\
&= \frac{2\sigma^2}{\sqrt{2\pi}} \left(I_2 \left(\frac{s_M}{\sigma} \right) - I_2(0) \right) + \frac{2s_M^2}{\sqrt{2\pi}} \left(\lim_{x \to +\infty} I_0(x) - I_0 \left(\frac{s_M}{\sigma} \right) \right) \\
&= 2\sigma^2 \left(\frac{1}{2} - \frac{s_M}{\sigma\sqrt{2\pi}} e^{-\frac{s_M^2}{2\sigma^2}} - \left(1 - \frac{s_M^2}{\sigma^2} \right) Q \left(\frac{s_M}{\sigma} \right) \right). \tag{B.18}
\end{aligned}
$$

The total power associated to the IMP of order $2\gamma + 1$ depends on coefficients $\nu_{2\gamma+1}$ that can be calculated by replacing (3.3) in (B.5), thus obtaining

$$
\begin{aligned}
\nu_{2\gamma+1} &= \sigma^2 \int_0^{s_M/\sigma} x H_{2\gamma+1} \left(\frac{x}{\sqrt{2}} \right) e^{-\frac{x^2}{2}} \, dx \\
&\quad + \sigma s_M \int_{s_M/\sigma}^{+\infty} H_{2\gamma+1} \left(\frac{x}{\sqrt{2}} \right) e^{-\frac{x^2}{2}} \, dx. \tag{B.19}
\end{aligned}
$$

From (B.7), we get

$$
\begin{aligned}
\nu_{2\gamma+1} &= \sigma^2 \sum_{m=0}^{\gamma} h_{\gamma,m} \int_0^{s_M/\sigma} x^{2m+2} \, e^{-\frac{x^2}{2}} \, dx \\
&\quad + \sigma s_M \sum_{m=0}^{\gamma} h_{\gamma,m} \int_{s_M/\sigma}^{+\infty} x^{2m+1} \, e^{-\frac{x^2}{2}} \, dx, \tag{B.20}
\end{aligned}
$$

and using (B.13) and (B.14), the above expression becomes

$$
\begin{aligned}
\nu_{2\gamma+1} &= \sum_{m=0}^{\gamma} h_{\gamma,m} \left[\sigma^2 \left(I_{2m+2} \left(\frac{s_M}{\sigma} \right) - I_{2m+2}(0) \right) \right. \\
&\qquad\qquad \left. + \sigma s_M \left(\lim_{x \to +\infty} I_{2m+1}(x) - I_{2m+1} \left(\frac{s_M}{\sigma} \right) \right) \right] \\
&= \sum_{m=0}^{\gamma} h_{\gamma,m} \left[\sigma^2 I_{2m+2} \left(\frac{s_M}{\sigma} \right) - \sigma s_M I_{2m+1} \left(\frac{s_M}{\sigma} \right) \right], \tag{B.21}
\end{aligned}
$$

since $I_{2m+2}(0) = (2m+1)!! \, I_0(0) = 0$ and $\lim_{x \to +\infty} I_{2m+1}(x) = 0$.

Polynomial Cartesian Memoryless Nonlinearity

Let us considerer a Cartesian memoryless nonlinearity characterized by the odd function $g(x)$, which can be written as a power series of form (3.2), with $\beta_m = 0$, $m \geq M$, i.e.,

$$g(x) = \sum_{m=0}^{M} \beta_m x^{2m+1}. \tag{B.22}$$

In this case, α can be calculated by replacing (B.22) in (B.1), thus obtaining

$$
\begin{aligned}
\alpha &= \frac{1}{\sqrt{2\pi}\sigma^3} \int_{-\infty}^{+\infty} \left(\sum_{m=0}^{M} \beta_m x^{2m+2} \right) e^{-\frac{x^2}{2\sigma^2}} dx \\
&= \frac{2}{\sqrt{2\pi}} \sum_{m=0}^{M} \beta_m \sigma^{2m} \int_{0}^{+\infty} x^{2m+2} e^{-\frac{x^2}{2}} dx \\
&= \frac{2}{\sqrt{2\pi}} \sum_{m=0}^{M} \beta_m \sigma^{2m} M_{2m+2},
\end{aligned}
\tag{B.23}
$$

where

$$M_m = \int_{0}^{+\infty} x^m e^{-\frac{x^2}{2}} dx \tag{B.24}$$

can be obtained from (B.13) and (B.14) for odd and even values of n, respectively, (see [Zwi03])

$$M_{2n} = \int_{0}^{+\infty} x^{2n} e^{-\frac{x^2}{2}} dx = \frac{\sqrt{2\pi}}{2} (2n-1)!! \tag{B.25}$$

and

$$M_{2n+1} = \int_{0}^{+\infty} x^{2n+1} e^{-\frac{x^2}{2}} dx = 2^n \, n!. \tag{B.26}$$

In order to calculate the average power of the signal at the nonlinearity output, from (B.22), we can write

$$g^2(x) = \sum_{m=0}^{M} \sum_{m'=0}^{M} \beta_m \beta_{m'} x^{2(m+m')+2} = \sum_{m=0}^{2M} \zeta_m x^{2m+2}, \tag{B.27}$$

with

$$\zeta_m = \sum_{k=0}^{m} \beta_k \beta_{m-k}. \tag{B.28}$$

Hence, replacing (B.27) in (B.2), it is straightforward to obtain

$$
\begin{aligned}
P_{\text{out}} &= \frac{1}{\sqrt{2\pi\sigma^2}} \int_{-\infty}^{+\infty} \sum_{m=0}^{2M} \zeta_m x^{2m+2} e^{-\frac{x^2}{2\sigma^2}} dx \\
&= \frac{2}{\sqrt{2\pi}} \sum_{m=0}^{2M} \zeta_m \sigma^{2m+2} \int_0^{+\infty} x^{2m+2} e^{-\frac{x^2}{2}} dx \\
&= \frac{2}{\sqrt{2\pi}} \sum_{m=0}^{2M} \zeta_m \sigma^{2m+2} M_{2m+2}.
\end{aligned}
\tag{B.29}
$$

The total power associated with the IMP of order $2\gamma + 1$ is given by (B.4) and can be obtained by replacing (B.7) and (B.22) in (B.5)

$$
\begin{aligned}
\nu_{2\gamma+1} &= \sigma \sum_{m=0}^{M} \sum_{k=0}^{\gamma} \int_0^{+\infty} \beta_m (\sigma x)^{2m+1} h_{\gamma,k} x^{2k+1} e^{-\frac{x^2}{2}} dx \\
&= \sum_{m=0}^{M} \sum_{k=0}^{\gamma} \beta_m h_{\gamma,k} \sigma^{2m+2} M_{2m+2k+2}.
\end{aligned}
\tag{B.30}
$$

B.2 Polar Memoryless Nonlinearities

Let us consider a polar memoryless nonlinearity characterized by the functions $A(R)$ e $\Theta(R)$ (see Figure 3.5). In Section 3.4, we showed that the useful component scale factor α is given by

$$
\alpha = \frac{E[Rf(R)]}{E[R^2]} = \frac{1}{2\sigma^4} \int_0^{+\infty} R^2 f(R) e^{-\frac{R^2}{2\sigma^2}} dR,
\tag{B.31}
$$

the average power of the signal at the nonlinearity output is

$$
P_{\text{out}} = \frac{1}{2} E[f^2(R)] = \frac{1}{2\sigma^2} \int_0^{+\infty} R f^2(R) e^{-\frac{R^2}{2\sigma^2}} dR
\tag{B.32}
$$

and the total power associated with the IMP of order $2\gamma + 1$ is given by

$$
P_{2\gamma+1} = \frac{1}{4\sigma^6(\gamma+1)} \left| \int_0^{+\infty} R^2 f(R) e^{-\frac{R^2}{2\sigma^2}} L_\gamma^{(1)} \left(\frac{R^2}{2\sigma^2} \right) dR \right|^2,
\tag{B.33}
$$

which can be written as

$$
P_{2\gamma+1} = \frac{1}{4(\gamma+1)} \left| \int_0^{+\infty} R^2 f(\sigma R) e^{-R^2/2} L_\gamma^{(1)} \left(\frac{R^2}{2} \right) dR \right|^2.
\tag{B.34}
$$

Next expressions for α, P_{out} and $P_{2\gamma+1}$ are found for the cases of $f(R)$ being an ideal envelope clipping and when it can be expanded as a power series. We will use the fact that Laguerre polynomials can be expanded as

$$L_n^{(m)}(x) = \sum_{k=0}^{n} \frac{(-1)^k}{k!} \binom{n+m}{n-k} x^k,$$ (B.35)

which allows us to write

$$
\begin{aligned}
L_\gamma^{(1)}\left(\frac{R^2}{2}\right) &= \sum_{k=0}^{\gamma} \frac{(-1)^k}{k!} \binom{\gamma+1}{\gamma-k} \left(\frac{R^2}{2}\right)^k \\
&= \sum_{k=0}^{\gamma} \frac{(-1)^k 2^{-k}}{k!} \binom{\gamma+1}{\gamma-k} R^{2k} \\
&= \sum_{k=0}^{\gamma} l_{\gamma,k} R^{2k},
\end{aligned}
$$ (B.36)

with

$$l_{\gamma,k} = \frac{(-1)^k 2^{-k}}{k!} \binom{\gamma+1}{\gamma-k}.$$ (B.37)

Ideal Envelope Clipping

In this case, the nonlinear function is a real function expressed by (3.25)

$$f(R) = A(R)\, e^{j\Theta(R)} = \begin{cases} R, & R \le s_M \\ s_M, & R > s_M. \end{cases}$$ (B.38)

Substituting this expression into (B.31), the useful component scale factor α becomes

$$
\begin{aligned}
\alpha &= \frac{1}{2\sigma^4} \int_0^{s_M} R^3\, e^{-\frac{R^2}{2\sigma^2}}\, dR + \frac{1}{2\sigma^4} \int_{s_M}^{+\infty} s_M R^2\, e^{-\frac{R^2}{2\sigma^2}}\, dR \\
&= \frac{1}{2} \int_0^{s_M/\sigma} R^3\, e^{-\frac{R^2}{2}}\, dR + \frac{s_M}{2\sigma} \int_{s_M/\sigma}^{+\infty} R^2\, e^{-\frac{R^2}{2}}\, dR \\
&= \frac{1}{2}\left(I_3\left(\frac{s_M}{\sigma}\right) - I_3(0)\right) + \frac{s_M}{2\sigma}\left(\lim_{R\to+\infty} I_2(R) - I_2\left(\frac{s_M}{\sigma}\right)\right).
\end{aligned}
$$ (B.39)

From (B.13), we write

$$I_3(x) = -x^2\, e^{-\frac{x^2}{2}} - 2\, e^{-\frac{x^2}{2}},$$ (B.40)

which, together with (B.16), allows us to write

$$\alpha = 1 - e^{-\frac{s_M^2}{2\sigma^2}} + \frac{\sqrt{2\pi}\, s_M}{2\sigma} Q\left(\frac{s_M}{\sigma}\right).$$ (B.41)

Replacing (B.38) in (B.32), we get

$$
\begin{aligned}
P_{\text{out}} &= \frac{1}{2\sigma^2} \int_0^{s_M} R^3 \, e^{-\frac{R^2}{2\sigma^2}} \, dR + \frac{s_M^2}{2\sigma^2} \int_{s_M}^{+\infty} R \, e^{-\frac{R^2}{2\sigma^2}} \, dR \\
&= \frac{\sigma^2}{2} \int_0^{s_M/\sigma} R^3 \, e^{-\frac{R^2}{2}} \, dR + \frac{s_M^2}{2} \int_{s_M/\sigma}^{+\infty} R \, e^{-\frac{R^2}{2}} \, dR \\
&= \frac{\sigma^2}{2} \left(I_3 \left(\frac{s_M}{\sigma} \right) - I_3(0) \right) + \frac{s_M^2}{2} \left(\lim_{R \to +\infty} I_1(R) - I_1 \left(\frac{s_M}{\sigma} \right) \right) \\
&= \sigma^2 \left(1 - e^{-\frac{s_M^2}{2\sigma^2}} \right).
\end{aligned} \tag{B.42}
$$

Using (B.38) and (B.36), expression (B.34) becomes

$$
\begin{aligned}
P_{2\gamma+1} &= \frac{1}{4(\gamma+1)} \left| \sigma \int_0^{s_M/\sigma} R^3 L_\gamma^{(1)} \left(\frac{R^2}{2} \right) e^{-\frac{R^2}{2}} \, dR \right. \\
&\qquad \left. + s_M \int_{s_M/\sigma}^{+\infty} R^2 L_\gamma^{(1)} \left(\frac{R^2}{2} \right) e^{-\frac{R^2}{2}} \, dR \right|^2 \\
&= \frac{1}{4(\gamma+1)} \left(\sigma \sum_{k=0}^{\gamma} l_{\gamma,k} \int_0^{s_M/\sigma} R^{2k+3} \, e^{-\frac{R^2}{2}} \, dR \right. \\
&\qquad \left. + s_M \sum_{k=0}^{\gamma} l_{\gamma,k} \int_{s_M/\sigma}^{+\infty} R^{2k+2} \, e^{-\frac{R^2}{2}} \, dR \right)^2 \\
&= \frac{1}{4(\gamma+1)} \left(\sum_{k=0}^{\gamma} l_{\gamma,k} \left(\sigma 2^{k+1} \int_0^{s_M^2/(2\sigma^2)} x^{k+1} \, e^{-x} \, dx \right. \right. \\
&\qquad \left. \left. + s_M \int_{s_M/\sigma}^{+\infty} R^{2k+2} \, e^{-\frac{R^2}{2}} \, dR \right) \right)^2. \tag{B.43}
\end{aligned}
$$

Using $I_n(x)$ given by (B.14) and defining $S_n(x)$ as

$$
S_n(x) = \int x^n \, e^{-x} \, dx, \tag{B.44}
$$

which is given by (see [GR07])

$$
\begin{aligned}
S_n(x) &= -e^{-x} \sum_{k=0}^{n} \frac{n!}{(n-k)!} x^{n-k} \\
&= -e^{-x} (x^n + n x^{n-1} + n(n-1)x^{n-2} + \ldots + n!), \tag{B.45}
\end{aligned}
$$

the IMPs can be written as

$$P_{2\gamma+1} = \frac{1}{4(\gamma+1)} \left[\sum_{k=0}^{\gamma} l_{\gamma,k} \left(2^{k+1} \sigma \left(S_{k+1} \left(\frac{s_M^2}{2\sigma^2} \right) - S_{k+1}(0) \right) \right. \right.$$

$$\left. \left. + s_M \left(\lim_{R\to\infty} I_{2k+2}(R) - I_{2k+2} \left(\frac{s_M}{\sigma} \right) \right) \right) \right]^2$$

$$= \frac{1}{4(\gamma+1)} \left[\sum_{k=0}^{\gamma} l_{\gamma,k} \left(2^{k+1} \sigma S_{k+1} \left(\frac{s_M^2}{2\sigma^2} \right) \right. \right.$$

$$\left. \left. -2^{k+1} \sigma(k+1)! - s_M I_{2k+2} \left(\frac{s_M}{\sigma} \right) \right) \right]^2 . \quad \text{(B.46)}$$

Polynomial Polar Memoryless Nonlinearities

Let us consider a polar memoryless nonlinearity whose characteristic can be expanded as a series of the form (3.23), with $\beta_m = 0, m \geq M$, i.e.,

$$f(R) = A(R) e^{j\Theta(R)} = \sum_{m=0}^{M} \beta_m R^{2m+1}. \quad \text{(B.47)}$$

In this case, α can be obtained substituting (B.47) in (B.31), yielding

$$\alpha = \frac{1}{2\sigma^4} \int_0^{+\infty} \sum_{m=0}^{M} \beta_m R^{2m+3} e^{-\frac{R^2}{2\sigma^2}} dR$$

$$= \frac{1}{2\sigma^4} \sum_{m=0}^{M} \beta_m \sigma^{2m+4} \int_0^{+\infty} R^{2m+3} e^{-\frac{R^2}{2}} dR. \quad \text{(B.48)}$$

Using (B.25), this reduces to

$$\alpha = \sum_{m=0}^{M} \beta_m 2^m (m+1)! \, \sigma^{2m}. \quad \text{(B.49)}$$

Since $f(R) = A(R) e^{j\Theta(R)} = A_I(R) + jA_Q(R)$, with $A_I(R) = A(R)\cos(\Theta(R))$ and $A_Q(R) = A(R)\sin(\Theta(R))$, we can write

$$A^2(R) = A_I^2(R) + A_Q^2(R). \quad \text{(B.50)}$$

From (B.47), we get

$$A_I(R) = \sum_{m=0}^{M} \text{Re}\{\beta_m\} R^{2m+1} \quad \text{(B.51)}$$

and

$$A_Q(R) = \sum_{m=0}^{M} \text{Im}\{\beta_m\} R^{2m+1}. \tag{B.52}$$

Using a procedure analogous to (B.27) and (B.28) for $A_I^2(R)$ and $A_Q^2(R)$, we obtain

$$A^2(R) = \sum_{m=0}^{2M} \zeta_m R^{2m+2}, \tag{B.53}$$

with

$$\zeta_m = \sum_{k=0}^{m} \left(\text{Re}\{\beta_k\} \text{Re}\{\beta_{m-k}\} + \text{Im}\{\beta_k\} \text{Im}\{\beta_{m-k}\} \right). \tag{B.54}$$

Replacing (B.53) in (B.32) and using (B.25), we get

$$
\begin{aligned}
P_{\text{out}} &= \frac{1}{2\sigma^2} \int_0^{+\infty} \sum_{m=0}^{2M} \zeta_m R^{2m+3} e^{-\frac{R^2}{2\sigma^2}} dR \\
&= \frac{1}{2\sigma^2} \sum_{m=0}^{2M} \zeta_m \sigma^{2m+4} \int_0^{+\infty} R^{2m+3} e^{-\frac{R^2}{2}} dR \\
&= \frac{1}{2\sigma^2} \sum_{m=0}^{2M} \zeta_m 2^{m+1} (m+1)! \, \sigma^{2m+4} \\
&= \sum_{m=0}^{2M} \zeta_m 2^m (m+1)! \, \sigma^{2m+2}.
\end{aligned} \tag{B.55}
$$

Using (B.47) and (B.36) in (B.57), we obtain

$$
\begin{aligned}
P_{2\gamma+1} &= \frac{1}{4(\gamma+1)} \left| \sum_{m=0}^{M} \beta_m \sigma^{2m+1} \int_0^{+\infty} R^2 R^{2m+1} e^{-R^2/2} L_\gamma^{(1)} \left(\frac{R^2}{2} \right) dR \right|^2 \\
&= \frac{1}{4(\gamma+1)} \left(\sum_{m=0}^{M} \beta_m \sigma^{2m+1} \int_0^{+\infty} R^{2m+3} e^{-R^2/2} \sum_{k=0}^{\gamma} l_{\gamma,k} R^{2k} dR \right)^2 \\
&= \frac{1}{4(\gamma+1)} \left(\sum_{m=0}^{M} \beta_m \sigma^{2m+1} \sum_{k=0}^{\gamma} l_{\gamma,k} \int_0^{+\infty} R^{2m+2k+3} e^{-R^2/2} dR \right)^2.
\end{aligned} \tag{B.56}
$$

Using (B.25), this becomes

$$
P_{2\gamma+1} = \frac{1}{4(\gamma+1)} \left(\sum_{m=0}^{M} \beta_m \sigma^{2m+1} \sum_{k=0}^{\gamma} l_{\gamma,k} 2^{m+k+1} (m+k+1)! \right)^2
$$

$$
= \frac{1}{\gamma+1} \left(\sum_{m=0}^{M} \beta_m 2^m \sigma^{2m+1} \sum_{k=0}^{\gamma} \frac{(-1)^k}{k!} \binom{\gamma+1}{\gamma-k} (m+k+1)! \right)^2.
$$

$$(B.57)$$

Polar Memoryless Nonlinearities of the Form R^{2p+1}

We now particularize the results obtained for a polar memoryless nonlinearity whose characteristic can be written as (B.47) for the case

$$
\beta_m = \begin{cases} 1, & m = p \\ 0, & m \neq p, \end{cases}
$$

$$(B.58)$$

i.e.,

$$
f^{(2p+1)}(R) = R^{2p+1}.
$$

$$(B.59)$$

In this case, (B.49) simply becomes

$$
\alpha^{(2p+1)} = 2^p (p+1)! \, \sigma^{2p},
$$

$$(B.60)$$

the average power of the signal at the nonlinearity output given by (B.55) reduces to

$$
P_{\text{out}}^{(2p+1)} = \zeta_{2p} 2^{2p} (2p+1)! \, \sigma^{4p+2} = 2^{2p} (2p+1)! \, \sigma^{4p+2}.
$$

$$(B.61)$$

The IMPs can be written as

$$
P_{2\gamma+1}^{(2p+1)} = \frac{1}{\gamma+1} \left(2^p \sigma^{2p+1} \sum_{k=0}^{\gamma} \frac{(-1)^k}{k!} \binom{\gamma+1}{\gamma-k} (p+k+1)! \right)^2
$$

$$
= \frac{1}{\gamma+1} \left(2^p \sigma^{2p+1} \sum_{k=0}^{\gamma} \frac{(-1)^k}{k!} \frac{(\gamma+1)!(p+k+1)!}{(k+1)!(\gamma-k)!} \right)^2. \quad (B.62)
$$

It can be shown that

$$
\sum_{k=0}^{\gamma} (-1)^k \frac{(\gamma+1)!(k+p+1)!}{k!(k+1)!(\gamma-k)!} = (-1)^\gamma \frac{p!}{\gamma!} (p+1)p \ldots (p-\gamma+1)
$$

$$
= (-1)^\gamma \frac{p!}{\gamma!} \prod_{p'=0}^{\gamma} (p+1-p')
$$

$$
= (-1)^\gamma \frac{p!(p+1)!}{\gamma!(p-\gamma)!}, \qquad (B.63)
$$

hence, the IMPs can be obtained from

$$
\begin{aligned}
P_{2\gamma+1}^{(2p+1)} &= \frac{1}{\gamma+1}\left(2^p \sigma^{2p+1}(-1)^\gamma \frac{p!(p+1)!}{\gamma!(p-\gamma)!}\right)^2 \\
&= \frac{1}{\gamma+1}\left(\frac{p!(p+1)!}{\gamma!(p-\gamma)!}\right)^2 2^{2p}\sigma^{4p+2}.
\end{aligned}
\tag{B.64}
$$

In the particular case of $\gamma = p$, we simply get

$$
P_{2p+1}^{(2p+1)} = p!(p+1)!\, 2^{2p}\sigma^{4p+2}.
\tag{B.65}
$$

These expressions can be rewritten in order to obtain α, P_{out} and $P_{2\gamma+1}$ for a given $f^{(2p+1)}(R) = R^{(2p+1)}$ from the 'previous' function $f^{(2p-1)}(R) = R^{(2p-1)}$, as follows

$$
\alpha^{(2p+1)} = (p+1)\, 2\sigma^2\, \alpha^{(2p-1)},
\tag{B.66}
$$

$$
P_{\text{out}}^{(2p+1)} = (2p+1)\, 4\sigma^4\, P_{\text{out}}^{(2p-1)}
\tag{B.67}
$$

and

$$
P_{2\gamma+1}^{(2p+1)} =
\begin{cases}
\dfrac{p^2(p+1)^2}{(p-\gamma)^2}\, 4\sigma^4\, P_{2\gamma+1}^{(2p-1)}, & \gamma < p \\[2mm]
p(p+1)\, 4\sigma^4\, P_{2p-1}^{(2p-1)}, & \gamma = p.
\end{cases}
\tag{B.68}
$$

Appendix C

Exact Characterization of a Polar Memoryless Nonlinearity with Characteristic $g(R) = R^5$

In this appendix, we present the exact characterization for a polar memoryless nonlinear device with characteristic of order 5, i.e., $\beta_m = 0$ for $m \geq 3$. Without loss of generality, we will assume $g(x) = x^5$, i.e., $\beta_0 = 0$, $\beta_1 = 0$ and $\beta_2 = 1$. Therefore,

$$
s_n^{(5)} = s_n s_n^* s_n s_n^* s_n
$$

$$
= \frac{1}{\sqrt{N^5}} \sum_{k_1=0}^{N-1} \sum_{k_2=0}^{N-1} \sum_{k_3=0}^{N-1} \sum_{k_4=0}^{N-1} \sum_{k_5=0}^{N-1} S_{k_1} S_{k_2}^* S_{k_3} S_{k_4}^* S_{k_5} e^{j2\pi(k_1 - k_2 + k_3 - k_4 + k_5)n/N'}
$$

$$
= \frac{1}{\sqrt{N^5}} \sum_{k^{(5)} \in \mathcal{K}^{(5)}} S_{k_1} S_{k_2}^* S_{k_3} S_{k_4}^* S_{k_5} e^{j2\pi(k_1 - k_2 + k_3 - k_4 + k_5)n/N'}, \quad \text{(C.1)}
$$

with $k^{(5)} = (k_1, k_2, k_3, k_4, k_5)$ and $\mathcal{K}^{(5)} = \mathcal{K}^{(1)} \times \mathcal{K}^{(1)} \times \mathcal{K}^{(1)} \times \mathcal{K}^{(1)} \times \mathcal{K}^{(1)}$.

Having in mind determining the multiplicity of each subcarrier, we note that $k_1 - k_2 + k_3 - k_4 + k_5 = k$ in the following cases

$$k = k_1 \qquad \begin{cases} k_2 = k_3 = k', k_4 = k_5 = k'' \\ k_4 = k_3 = k', k_2 = k_5 = k'' \end{cases} \qquad \text{(C.2a)}$$

$$k = k_3 \qquad \begin{cases} k_2 = k_1 = k', k_4 = k_5 = k'' \\ k_4 = k_1 = k', k_2 = k_5 = k'' \end{cases} \qquad \text{(C.2b)}$$

$$k = k_5 \qquad \begin{cases} k_2 = k_3 = k', k_4 = k_1 = k'' \\ k_4 = k_3 = k', k_2 = k_1 = k''. \end{cases} \qquad \text{(C.2c)}$$

Thus, we define the sets

$$\mathcal{K}_1^{(5)} = \{(k_1, k_2, k_3, k_4, k_5) \in \mathcal{K}^{(5)} : k_2 = k_1 \wedge k_4 = k_3\} \qquad \text{(C.3a)}$$

$$\mathcal{K}_2^{(5)} = \{(k_1, k_2, k_3, k_4, k_5) \in \mathcal{K}^{(5)} : k_2 = k_1 \wedge k_4 = k_5\} \qquad \text{(C.3b)}$$

$$\mathcal{K}_3^{(5)} = \{(k_1, k_2, k_3, k_4, k_5) \in \mathcal{K}^{(5)} : k_2 = k_3 \wedge k_4 = k_1\} \qquad \text{(C.3c)}$$

$$\mathcal{K}_4^{(5)} = \{(k_1, k_2, k_3, k_4, k_5) \in \mathcal{K}^{(5)} : k_2 = k_3 \wedge k_4 = k_5\} \qquad \text{(C.3d)}$$

$$\mathcal{K}_5^{(5)} = \{(k_1, k_2, k_3, k_4, k_5) \in \mathcal{K}^{(5)} : k_2 = k_5 \wedge k_4 = k_1\} \qquad \text{(C.3e)}$$

$$\mathcal{K}_6^{(5)} = \{(k_1, k_2, k_3, k_4, k_5) \in \mathcal{K}^{(5)} : k_2 = k_5 \wedge k_4 = k_3\}. \qquad \text{(C.3f)}$$

Clearly, $|\mathcal{K}^{(5)}| = N^5$ and $|\mathcal{K}_1^{(5)}| = |\mathcal{K}_2^{(5)}| = |\mathcal{K}_3^{(5)}| = |\mathcal{K}_4^{(5)}| = |\mathcal{K}_5^{(5)}| = |\mathcal{K}_6^{(5)}| = N^2$. The multiplicity of the useful component is

$$M_k^{(5 \to 1)} = \begin{cases} |\mathcal{U}^{(5)}|, & 0 \leq k \leq N - 1 \\ 0, & \text{otherwise,} \end{cases} \qquad \text{(C.4)}$$

with

$$\mathcal{U}^{(5)} = \bigcup_{m=1}^{6} \mathcal{K}_m^{(5)}. \qquad \text{(C.5)}$$

The cardinal of the set $\mathcal{U}^{(5)}$ can be obtained from

$$\left| \bigcup_{m=1}^{6} \mathcal{K}_m^{(5)} \right| = \sum_{m=1}^{6} |\mathcal{K}_m^{(5)}| + \sum_{m=2}^{6} (-1)^{m-1} \mathcal{I}_m^{(5)}$$

$$= \sum_{m=1}^{6} (-1)^{m-1} \mathcal{I}_m^{(5)}, \qquad \text{(C.6)}$$

with

$$\mathcal{I}_m^{(5)} = \sum_{\{i_1, \ldots, i_m\} \in \mathcal{C}_m^{(5)}} \left| \bigcap_{l=1}^{m} \mathcal{K}_{i_l}^{(5)} \right|, \qquad \text{(C.7)}$$

where $\mathcal{C}_m^{(5)}$ represents the set of all subsets of $\{1, 2, \ldots, 6\}$ that contain m elements. Clearly, $|\mathcal{C}_m^{(5)}| = \binom{6}{m}$, hence, for $m = 2$, the number of subsets is $\binom{6}{2} = 15$. It can be shown that of these, 9 have cardinal N and the remaining have cardinal 1, which means that

$$\mathcal{I}_2^{(5)} = 9N + 6. \tag{C.8}$$

For $m = 3$, 4, 5 and 6, all subsets have cardinal 1, hence,

$$\mathcal{I}_m^{(5)} = \binom{6}{m} \tag{C.9}$$

and

$$\begin{aligned}
\left|\mathcal{U}^{(5)}\right| &= \mathcal{I}_1^{(5)} - \mathcal{I}_2^{(5)} + \mathcal{I}_3^{(5)} - \mathcal{I}_4^{(5)} + \mathcal{I}_5^{(5)} - \mathcal{I}_6^{(5)} \\
&= 6N^2 - (9N - 6) + \binom{6}{3} - \binom{6}{4} + \binom{6}{5} - \binom{6}{6} \\
&= 6N^2 - 9N - 6 + 20 - 15 + 6 - 1 = 6N^2 - 9N + 4, \quad \text{(C.10)}
\end{aligned}$$

which means that

$$M_k^{(5\to1)} = (6N^2 - 9N + 4)M_k^{(1)}, \tag{C.11}$$

and the sum of the multiplicities of the useful and self-interference samples are

$$\sum_{k=0}^{N-1} M_k^{(5\to1)} = 6N^3 - 9N^2 + 4N \tag{C.12}$$

and

$$\sum_{k=-2N+2}^{3N-3} M_k^{(5)} - \sum_{k=0}^{N-1} M_k^{(5\to1)} = N^5 - 6N^3 + 9N^2 - 4N, \tag{C.13}$$

respectively. This means that we can write $s_n^{(5)}$ as a sum of a useful and a self-interference components

$$\begin{aligned}
s_n^{(5)} &= (6N^2 - 9N + 4)(E[|S_k|^2])^2 \frac{1}{\sqrt{N^5}} \sum_{k \in \mathcal{K}^{(1)}} S_k e^{j2\pi kn/N'} \\
&\quad + \frac{1}{\sqrt{N^5}} \sum_{k^{(5)} \in \mathcal{K}^{(5)} \setminus \mathcal{U}^{(5)}} S_{k_1} S_{k_2}^* S_{k_3} S_{k_4}^* S_{k_5} e^{j2\pi(k_1 - k_2 + k_3 - k_4 + k_5)n/N'} \\
&= \alpha^{(5)} s_n + d_n^{(5)}, \tag{C.14}
\end{aligned}$$

with

$$\alpha^{(5)} = \frac{6N^2 - 9N + 4}{N^2}(E[|S_k|^2])^2$$

$$= \left(6 - \frac{9}{N} + \frac{4}{N^2}\right)(E[|S_k|^2])^2 \qquad (C.15)$$

and

$$d_n^{(5)} = \frac{1}{\sqrt{N^5}} \sum_{k^{(5)} \in \mathcal{K}^{(5)} \setminus \mathcal{U}^{(5)}} S_{k_1} S_{k_2}^* S_{k_3} S_{k_4}^* S_{k_5} e^{j2\pi(k_1 - k_2 + k_3 - k_4 + k_5)n/N'}.$$

$$(C.16)$$

Consider the following cases

$$\begin{aligned}
k_2 &= k_1 = k \\
k_2 &= k_3 = k \\
k_2 &= k_5 = k \\
k_4 &= k_1 = k \\
k_4 &= k_3 = k \\
k_4 &= k_5 = k.
\end{aligned} \qquad (C.17)$$

Clearly, this allows us to write the self-interference samples as a sum of two components

$$\begin{aligned}
d_n^{(5)} &= \frac{1}{\sqrt{N^5}} \sum_k \sum_k |S_k|^2 \sum_{k'} \sum_{k''} \sum_{k'''} S_{k'} S_{k''}^* S_{k'''} e^{j2\pi(k' - k'' + k''')n/N'} \\
&+ \frac{1}{\sqrt{N^5}} \sum_{k_1} \sum_{k_2} \sum_{k_3} \sum_{k_4} \sum_{k_5} S_{k_1} S_{k_2}^* S_{k_3} S_{k_4}^* S_{k_5} e^{j2\pi(k_1 - k_2 + k_3 - k_4 + k_5)n/N'} \\
&= \frac{1}{N} \sum_k M_k^{(5 \to 3)} E[|S_k|^2] \underbrace{\frac{1}{\sqrt{N^3}} \sum_{k'} \sum_{k''} \sum_{k'''} S_{k'} S_{k''}^* S_{k'''} e^{j2\pi(k' - k'' + k''')n/N'}}_{\approx d_n^{(3)}} \\
&+ \frac{1}{\sqrt{N^5}} \sum_{k_1} \sum_{k_2} \sum_{k_3} \sum_{k_4} \sum_{k_5} S_{k_1} S_{k_2}^* S_{k_3} S_{k_4}^* S_{k_5} e^{j2\pi(k_1 - k_2 + k_3 - k_4 + k_5)n/N'}.
\end{aligned}$$

$$(C.18)$$

It can be shown that

$$M_k^{(5 \to 3)} \approx (6N - 9)M_k^{(3 \to 3)} + (N - 2)M_k^{(1)} \qquad (C.19)$$

and

$$\sum_{k=-2N+2}^{3N-3} M_k^{(5 \to 3)} = (6N - 9)(N^3 - 2N^2 + N) + (N - 2)N + N^2$$

$$= 6N^4 - 21N^3 + 26N^2 - 11N. \qquad (C.20)$$

Therefore,

$$\sum_{k=-2N+2}^{3N-3} M_k^{(5\to5)} = \sum_{k=-2N+2}^{3N-3} M_k^{(5)} - \sum_{k=-2N+2}^{3N-3} M_k^{(5\to3)} - \sum_{k=0}^{N-1} M_k^{(5\to1)}$$

$$= N^5 - (6N^3 - 9N^2 + 4N)$$
$$- (6N^4 - 21N^3 + 26N^2 - 11N)$$
$$= N^5 - 6N^4 + 15N^3 - 17N^2 + 7N. \qquad (C.21)$$

The power of the useful component is

$$S^{(5)} = P_1^{(5)} = \frac{(\alpha^{(5)})^2}{2} E[|S_k|^2], \qquad (C.22)$$

and the power of the self-interfering components are

$$P_3^{(5)} = \frac{1}{N^2}(6N - 9)^2 \, (E[|S_k|^2])^2 P_3^{(3)}$$

$$= \left(6 - \frac{9}{N}\right)^2 \left(1 - \frac{2}{N} + \frac{1}{N^2}\right) (E[|S_k|^2])^5 \qquad (C.23)$$

and

$$P_5^{(5)} = 3! \, 2! \, \frac{N^5 - 6N^4 + 15N^3 - 17N^2 + 7N}{N^5} \frac{(E[|S_k|^2])^5}{2}$$

$$= 6\left(1 - \frac{6}{N} + \frac{15}{N^2} - \frac{17}{N^3} + \frac{7}{N^4}\right)(E[|S_k|^2])^5, \qquad (C.24)$$

where the factors 3! and 2! are the number of times we can get repetitions of the values taken by k_1, k_3 and k_5, and k_2 and k_4, respectively. The total power of the self-interference samples is $I^{(5)} = P_3^{(5)} + P_5^{(5)}$.

Generalization

Previous results can be generalized for large values of N noticing that, if $\gamma < p$,

$$M_k^{(2p+1\to2\gamma+1)} \approx \binom{p}{p-\gamma} P(p+1, p-\gamma) N^{p+\gamma}$$

$$= \frac{p!}{\gamma!(p-\gamma)!} \frac{(p+1)!}{(\gamma+1)!} N^{p+\gamma}, \qquad (C.25)$$

where $P(n, r) = n!/(n-r)!$ denotes 'permutation of n elements r by r'. For $\gamma = p$, we have

$$M_k^{(2p+1\to2p+1)} \approx M_k^{(2p+1)} - \sum_{l=0}^{p-1} M_k^{(2p+1\to2l+1)}. \qquad (C.26)$$

Thus, the power of the useful and self-interfering components can be approximated by

$$
\begin{aligned}
P_{2\gamma+1}^{(2p+1)} &\approx \frac{1}{N^{2p+1}}\gamma!(\gamma+1)!\left(\sum_k M_k^{(2p+1\to2\gamma+1)}\right)^2\frac{(E[|S_k|^2])^{2p+1}}{2N^{2\gamma+1}} \\
&= \frac{1}{N^{2p+1}}\gamma!(\gamma+1)!\left(\frac{p!}{\gamma!(p-\gamma)!}\frac{(p+1)!}{(\gamma+1)!}N^{p+\gamma+1}\right)^2 \\
&\quad\times\frac{(E[|S_k|^2])^{2p+1}}{2N^{2\gamma+1}} \\
&= \frac{1}{N^{2p+1}}\frac{1}{\gamma!(\gamma+1)!}\left(\frac{p!(p+1)!}{(p-\gamma)!}\right)^2 N^{2p+2\gamma+2}\frac{(E[|S_k|^2])^{2p+1}}{2N^{2\gamma+1}} \\
&= \frac{1}{2}\frac{1}{\gamma!(\gamma+1)!}\left(\frac{p!(p+1)!}{(p-\gamma)!}\right)^2 (E[|S_k|^2])^{2p+1}.
\end{aligned}
\tag{C.27}
$$

For $p = \gamma$, we have

$$
\begin{aligned}
P_{2p+1}^{(2p+1)} &\approx \frac{1}{N^{2p+1}}p!(p+1)!\left(\sum_k M_k^{(2p+1\to2p+1)}\right)^2\frac{(E[|S_k|^2])^{2p+1}}{2N^{2p+1}} \\
&= \frac{1}{N^{2p+1}}p!(p+1)!\left(N^{2p+1}-\sum_{p=0}^{p-1}\sum_k M_k^{(2p+1\to2p+1)}\right)^2 \\
&\quad\times\frac{(E[|S_k|^2])^{2p+1}}{2N^{2p+1}} \\
&\approx \frac{1}{N^{2p+1}}p!(p+1)!\left(N^{2p+1}\right)^2\frac{(E[|S_k|^2])^{2p+1}}{2N^{2p+1}} \\
&\approx \frac{p!(p+1)!}{2}(E[|S_k|^2])^{2p+1}.
\end{aligned}
\tag{C.28}
$$

Appendix D

Approximated Symbol Error Rate for Cross QAM Constellations

In this appendix, we present a simple approach to the calculus of the symbol energy and an approximated Symbol Error Rate (SER) for cross Quadrature Amplitude Modulation (QAM) constellations.

Cross constellations are QAM constellations with 2^{2m+1} signal points, for $m \geq 2$. They are obtained from a square with $3 \times 2^{m-1}$ side points, to which four square corners are cut, with 2^{m-2} side points, thus obtaining a constellation with the form of a cross. It is always possible to build such constellations, since it can easily be proven by induction that the relation $2^{2m+1} = (3 \times 2^{m-1})^2 - 4 \times (2^{m-2})^2$ holds for $m \geq 2$. Signal points coordinates are $\pm a$, $\pm 3a$, $\pm 5a$, ..., so minimum Euclidean distance between them is $d = 2a$. Figures D.1 and D.2 show the first quadrants from 128 QAM and 512 QAM constellations, respectively.

Cross constellations can be seen as a set of square QAM constellations. Each quadrant of a cross constellation with 2^{2m+1} points can be seen as a square QAM constellation with $(2^{m-1})^2$ points, together with four smaller square QAM constellations, with $(2^{m-2})^2$ points each one, disposed as seen in Figure D.1. This decomposition is always possible, since $2^{2m+1} = 4((2^{m-1})^2 + 4 \times (2^{m-2})^2)$.

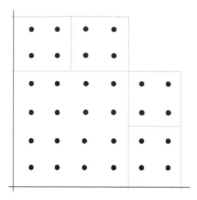

Figure D.1: Decomposition of the first quadrant of a 128 QAM constellation into smaller square QAM constellations.

Symbol Energy

We will use the decomposition of a cross QAM constellation into a set of square QAM constellations to calculate its symbol energy. It is well known that the symbol energy of a square QAM constellation with 2^{2m} points is

$$\varepsilon = \frac{2^{2m} - 1}{6} d^2. \tag{D.1}$$

The square subconstellation with 2^{2m-2} points has center $X_1 = x_1 + jy_1$. The minimum distance between points is $2a$, and the first point is $a + aj$, so $x_1 = y_1 = a + (2^{m-2} - 1)2a + a = 2^{m-1}a$ and $X_1 = 2^{m-1}a + j2^{m-1}a$. The energy of a square QAM constellation with 2^{2m-2} points is $(2^{2m-2} - 1)d^2/6$. This subconstellation is not centered at the origin, so its energy depends on its center X_1 and is given by

$$\varepsilon_1 = (2^{m-1}a)^2 + (2^{m-1}a)^2 + \frac{2^{2m-2} - 1}{6}(2a)^2$$

$$= \frac{2^{2m} - 1}{3} 2a^2. \tag{D.2}$$

One of the square subconstellations with 2^{2m-4} points has center $X_2 = x_2 + jy_2$. As before, minimum distance between points is $2a$, and the first point of this constellation has real part $2 \times 2^{m-1}a + a$, so the center of the constellation, X_2, has real and imaginary parts $x_2 = 2 \times 2^{m-1}a + a + (2^{m-3} - 1)2a + a = 5 \times 2^{m-2}a$ and $y_2 = a + (2^{m-3} - 1)2a + a = 2^{m-2}a$, respectively. A constellation with 2^{2m-4} points has energy $(2^{2m-4} - 1)d^2/6$, so, as this one is centered at X_2, its energy is

$$\varepsilon_2 = (5 \times 2^{m-2}a)^2 + (2^{m-2}a)^2 + \frac{2^{2m-4} - 1}{6}(2a)^2$$

$$= \frac{5 \times 2^{2m-1} - 1}{3}2a^2. \tag{D.3}$$

Another of the subconstellations is centered at $X_3 = x_3 + jy_3$, with $x_3 = x_2 = 5 \times 2^{m-2}a$. It is easily seen that $y_3 = 2 \times 2^{m-2}a + a + (2^{m-3} - 1)2a + a = 3 \times 2^{m-2}a$, so the constellation center is $X_3 = 5 \times 2^{m-2}a + 3 \times 2^{m-2}aj$. The number of constellation points is the same as before, but now centered at X_3, so its energy is given by

$$\varepsilon_3 = (5 \times 2^{m-2}a)^2 + (3 \times 2^{m-2}a)^2 + \frac{2^{2m-4} - 1}{6}(2a)^2$$

$$= \frac{13 \times 2^{2m-2} - 1}{3}2a^2. \tag{D.4}$$

The other two subconstellations with 2^{2m-4} points are symmetrical to these, so their energy is the same. The symbol energy for a cross constellation with 2^{2m+1} points is then given by

$$\varepsilon = \frac{1}{2}\left(\varepsilon_1 + \frac{2\varepsilon_2 + 2\varepsilon_3}{4}\right)$$

$$= \frac{31 \times 2^{2m-4} - 1}{6}d^2. \tag{D.5}$$

Symbol Error Rate

To determine the SER it is necessary to know the number of neighbours of each of the constellation points. Representing by A points with two nearest neighbours, by B and B_1 points with three nearest neighbours and by C and C_1 points with four nearest neighbours we obtain Figure D.2. Note that the decision boundaries of points of type A, B_1 and C_1 do not extend to the infinite, as shown in the figure. Due to this fact, the well-known exact results for the error probability (see [Car83]) for these type of points can only be used as approximations, and we can write

$$P(S|A) \approx 2Q(\eta) - (Q(\eta))^2 \tag{D.6a}$$

$$P(S|B) \approx P(S|B_1) \approx 3Q(\eta) - 2(Q(\eta))^2 \tag{D.6b}$$

$$P(S|C) \approx P(S|C_1) \approx 4Q(\eta) - 4(Q(\eta))^2, \tag{D.6c}$$

where $Q(\cdot)$ represents the well-known error function defined in (5.34), and η is given by (5.74). Using (D.5), η can be written for cross constel-

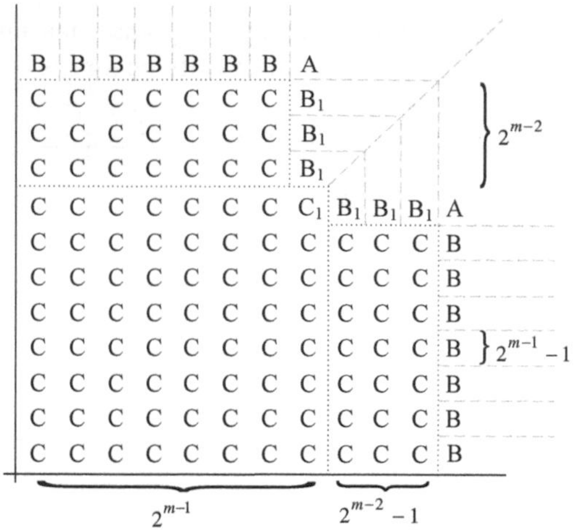

Figure D.2: Number of nearest neighbours and decision boundaries for a 512 QAM constellation.

lations as

$$\eta = \sqrt{\beta_{\text{cross}}\text{SNR}}, \tag{D.7}$$

where SNR denotes the channel output signal-to-noise ratio (defined in (5.75)) and

$$\beta_{\text{cross}} \triangleq \frac{3}{31 \times 2^{2m-4} - 1}. \tag{D.8}$$

It is clear that there are 2 points of type A in each quadrant of the constellation. The number of points B and B_1 is the same vertically or horizontally, and it is equal to $(2^{m-2} - 1) + (2^{m-1} - 1) = 2(3 \times 2^{m-2} - 2)$. Points of type C and C_1 form a square with side 2^{m-1} points and two rectangles with $(2^{m-2} - 1) \times (2^{m-1} - 1)$ points, so there is a total of $(2^{m-1})^2 + 2(2^{m-2} - 1)(2^{m-1} - 1)$ points of this type. It is easily checked that $2^{2m+1} = 4(2 + (3 \times 2^{m-1} - 4) + (2^{2m-1} - 3 \times 2^{m-1} + 2))$, with $m \geq 2$. Considering the number of points of each type, the SER of a cross constellation is approximately given by

$$P_s \approx \frac{4}{2^{2m+1}} \left[2P(S|A) + (3 \times 2^{m-1} - 4)P(S|B) \right.$$
$$\left. + (2^{2m-1} - 3 \times 2^{m-1} + 2)P(S|C) \right]. \tag{D.9}$$

Using (D.6), we can obtain the approximated SER from

$$P_s \approx \frac{2^{2m+3} - 3 \times 2^{m+1}}{2^{2m+1}} Q(\eta) + \frac{3 \times 2^{m+2} - 2^{2m+3} - 8}{2^{2m+1}} (Q(\eta))^2. \tag{D.10}$$

This expression can be written as

$$P_s \approx \alpha_1 Q(\eta) - \alpha_2 (Q(\eta))^2, \qquad \text{(D.11)}$$

where

$$\alpha_1 \triangleq 4 \left(1 - \frac{3}{2^{m+2}} \right) \qquad \text{(D.12)}$$

and

$$\alpha_2 \triangleq 4 \left(1 - \frac{3}{2^{m+1}} + \frac{1}{2^{2m}} \right). \qquad \text{(D.13)}$$

Neglecting the term depending on $(Q(\eta))^2$, we obtain the approximation

$$P_s \approx \alpha_1 Q(\eta). \qquad \text{(D.14)}$$

Figure D.3 shows the SER obtained using (D.11), (D.14), and the expressions presented in [BC07] and [LZB08] for 32, 128 and 512 cross QAM constellations. From the figure, it is clear that the results closely match and that for high values of the signal-to-noise ratio, there are no significant differences between approximations (D.11) and (D.14).

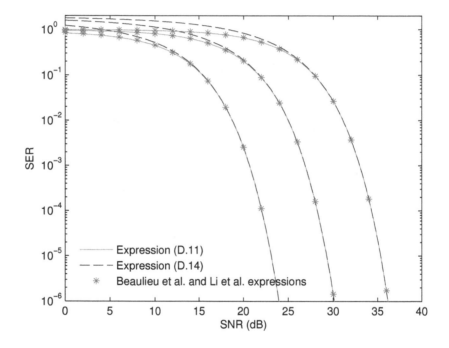

Figure D.3: Symbol error probability for 32 QAM, 128 QAM and 512 QAM.

References

[3GP06]   3GPPP TR 25.814. *3rd Generation Partnership Project: Technical Specification Group Radio Access Network; Physical Layers Aspects for Evolved UTRA*, 2006.

[AD06]   T. Araújo and R. Dinis. Efficient detection of zero-padded OFDM signals with large blocks. In *Proc. IASTED Signal and Image Processing (SIP'06)*, pages 358–361, Honolulu, Hawaii, USA, August 2006.

[AD07a]   T. Araújo and R. Dinis. Analytical evaluation and optimization of the analog-to-digital converter in software radio architectures. *IEEE Trans. Veh. Technol.*, 56(4):1964–1970, July 2007.

[AD07b]   T. Araújo and R. Dinis. Performance evaluation of quantization effects on multicarrier modulated signals. *IEEE Trans. Veh. Technol.*, 56(5):2922–2930, September 2007.

[AD08a]   T. Araújo and R. Dinis. Analytical evaluation of nonlinear distortion effects on OFDMA signals used in the downlink transmission. In *Proc. IEEE 5th International Symposium on Wireless Communication Systems (ISWCS'08)*, pages 31–36, Reykjavik, Iceland, October 2008.

[AD08b]   T. Araújo and R. Dinis. Analytical evaluation of nonlinear distortion effects on OFDMA uplink signals. In *Proc. IEEE 67th Vehicular Technology Conference (VTC'08 Spring)*, pages 938–942, Singapore, May 2008.

[AD09]   T. Araújo and R. Dinis. Loading techniques for OFDM systems with nonlinear distortion effects. In *Proc. 6th International Symposium on Wireless Communication Systems (ISWCS'09)*, pages 483–487, Siena, Italy, September 2009.

[AD10a]   T. Araújo and R. Dinis. Analytical evaluation of nonlinear effects on OFDMA signals. *IEEE Trans. Wireless Commun.*, 9(11):3472–3479, November 2010.

[AD10b]   T. Araújo and R. Dinis. On the accuracy of the Gaussian approximation for the evaluation of nonlinear effects in OFDM signals. In *Proc. IEEE 72nd Vehicular Technology Conference (VTC'10 Fall)*, Ottawa, Canada, September 2010.

[AD12a]   T. Araújo and R. Dinis. Loading techniques for OFDM systems with nonlinear distortion effects. *Transactions on Emerging Telecommunications Technologies*, 23(2):121–132, March 2012.

[AD12b]   T. Araújo and R. Dinis. On the accuracy of the Gaussian approximation for the evaluation of nonlinear effects in OFDM signals. *IEEE Trans. Commun.*, 60(2):346–351, February 2012.

[Ara12]   T. Araújo. *Analytical Evaluation of Nonlinear Distortion Effects in Multicarrier Signals*. PhD thesis, Instituto Superior Técnico, Universidade Técnica de Lisboa, 2012.

[Arm01]   J. Armstrong. New OFDM peak-to-average power reduction scheme. In *IEEE VTC'2001 (Spring)*, Rhodes, Greece, May 2001.

[Arm02]   J. Armstrong. Peak-to-average power reduction for OFDM by repeated clipping and frequency-domain filtering. *Electronics Letters*, 38(5):246–247, February 2002.

[AS72]   M. Abramowitz and I. Stegun. *Handbook of Mathematical Functions*. Dover Publications, New York, 1972.

[BB87]   S. Benedetto and E. Biglieri. *Digital Transmission*. Prentice Hall, New Jersey, 1987.

[BC00]   P. Banelli and S. Cacopardi. Theoretical analysis and performance of OFDM signals in nonlinear AWGN channels. *IEEE Trans. Commun.*, 48(3):430–441, March 2000.

[BC07]   N. C. Beaulieu and Y. Chen. Closed-form expressions for the exact symbol error probability of 32-cross-QAM in AWGN and in slow Nakagami fading. *IEEE Commun. Lett.*, 11(4):310–312, April 2007.

[Bin90]   J. A. C. Bingham. Multicarrier modulation for data transmission: An idea whose time has come. *IEEE Commun. Mag.*, 28(5):5–14, May 1990.

[Bin00]   J. A. C. Bingham. *ADSL, VDSL and Multicarrier Modulation.* John Wiley and Sons, Inc., 2000.

[Bla68]   N. M. Blachman. The uncorrelated output components of a nonlinearity. *IEEE Transactions on Information Theory*, 14(2):250–255, March 1968.

[BS02]   A. R. S. Bahai and B. R. Saltzberg. *Multi-Carrier Digital Communications: Theory and Applications of OFDM.* Kluwer Academic Publishers, Holland, Netherlands, 2002.

[BSGS02]   A. R. S. Bahai, M. Singh, A. J. Goldsmith, and B. R. Saltzberg. A new approach for evaluating clipping distortion in multicarrier systems. *IEEE J. Select. Areas Commun.*, 20(5):1037–1046, June 2002.

[Bus52]   J. Bussgang. Crosscorrelation function of amplitude-distorted gaussian signals. Technical Report 216, Research Laboratory of Electronics, Massachusetts Institute of Technology, Cambridge, Massachusetts, March 1952.

[Car83]   A. B. Carlson. *Communication Systems.* McGraw-Hill, New York, NY, 1983.

[Cim85]   L. Cimini, Jr. Analysis and simulation of a digital mobile channel using orthogonal frequency division multiplexing. *IEEE Trans. Commun.*, 33(7):665–675, July 1985.

[Cio91]   J. M. Cioffi. A multicarrier primer. Technical Report T1E1.4/91-159, Stanford University/Amati Communications Corporation, November 1991.

[CNS+05]   V. Chakravarthy, A. S. Nunez, J. P. Stephens, A. K. Shaw, and M. A. Temple. TDCS, OFDM, and MC-CDMA: A brief tutorial. *IEEE Commun. Mag.*, 43(9):S11–S16, September 2005.

[Cox74]   D. Cox.   Linear amplification with nonlinear compo-
             nents. *IEEE Trans. Commun.*, 22(12):1942–1945, Decem-
             ber 1974.

[CS99]   L. Cimini, Jr. and N. Sollenberger. Peak-to-average power
             reduction of an OFDM signal using partial transmit se-
             quences. *IEEE Commun. Lett.*, November 1999.

[CT65]   J. W. Cooley and J. W. Tukey. An algorithm for the ma-
             chine calculation of complex Fourier series. *Mathematics
             of Computation*, 19(90):297–301, April 1965.

[CWETM98]   A. Chini, Y. Wu, M. El-Tanany, and S. Mahmoud.
             Hardware nonlinearities in digital TV broadcasting using
             OFDM modulation. *IEEE Trans. Broadcast.*, 44(1):12–21,
             March 1998.

[Dar03]   D. Dardari.   Exact analysis of joint clipping and quan-
             tization effects in high speed WLAN receivers. In *Proc.
             IEEE International Conference on Communications 2003
             (ICC'03)*, volume 5, pages 3487–3492, Anchorage, Alaska,
             USA, May 2003.

[DFG+05]   E. Dahlman, P. Frenger, J. Guey, G. Klang, R. Ludwig,
             M. Meyer, N. Wiberg, and K. Zangi. A framework for fu-
             ture radio access. In *Proc. IEEE 61th Vehicular Technol-
             ogy Conference (VTC'05 Spring)*, volume 5, pages 2944–
             2948, Phoenix, Arizona, USA, May/June 2005.

[DG96a]   R. Dinis and A. Gusmão. CEPB-OFDM: A new technique
             for multicarrier tranmission with saturated power ampli-
             fiers. In *Proc. IEEE International Conference on Commu-
             nication Systems 1996 (ICCS'96)*, Singapore, November
             1996.

[DG96b]   R. Dinis and A. Gusmão.   Performance evaluation of
             OFDM transmission with convencional and two-branch
             combining power amplification schemes. In *Proc. IEEE
             Global Telecommunications Conference 1996 (GLOBE-
             COM'96)*, London, UK, November 1996.

[DG97]   R. Dinis and A. Gusmão. On the Gaussian aproxima-
             tion for evaluating nonlinear distortion effects in OFDM
             transmission. In *Proc. 4th Intl. Symp. Comm. Theory and
             Applications (ISCTA'97)*, Ambleside, UK, July 1997.

[DG98]   R. Dinis and A. Gusmão. Performance evaluation of a multicarrier modulation technique allowing strongly nonlinear amplification. In *IEEE ICC'98*, Atlanta, Georgia, USA, July 1998.

[DG00]   R. Dinis and A. Gusmão. A class of signal processing algorithms for good power/bandwidth tradeoffs with OFDM transmission. In *Proc. IEEE International Symposium on Information Theory 2000 (ISIT'00)*, page 216, Sorrento, Italy, June 2000.

[DG01]   R. Dinis and A. Gusmão. A new class of signal processing schemes for bandwidth-efficient OFDM transmission with low envelope fluctuation. In *Proc. IEEE VTS 53rd Vehicular Technology Conference Spring (VTC'01 Spring)*, volume 1, pages 658–662, Rhodes, Greece, May 2001.

[DG03a]   R. Dinis and A. Gusmão. An iterative technique for CEPB-OFDM transmission with low out-of-band radiation. In *IEEE VTC'03 (Fall)*, Orlando, Florida, USA, October 2003.

[DG03b]   R. Dinis and A. Gusmão. Performance evaluation of an iterative PMEPR-reducing technique for OFDM transmission. In *Proc. IEEE Global Telecommunications Conference 2003 (GLOBECOM'03)*, volume 1, pages 20–24, San Francisco, California, USA, December 2003.

[DG04]   R. Dinis and A. Gusmão. A class of nonlinear signal-processing schemes for bandwidth-efficient OFDM transmission with low envelope fluctuation. *IEEE Trans. Commun.*, 52(11):2009–2018, November 2004.

[DG08]   R. Dinis and A. Gusmão. Nonlinear signal processing schemes for OFDM modulations within conventional or LINC transmitter structures. *Eur. Trans. Telecomm.*, 19(3):257–271, April 2008.

[DHM57]   M. L. Doelz, E. T. Heald, and D. L. Martin. Binary data transmission techniques for linear systems. *Proceedings of the IRE*, 45(5):656–661, May 1957.

[Din01]   R. Dinis. *Técnicas Multiportadora para Rádio Móvel de Alto Débito*. PhD thesis, Instituto Superior Técnico, Universidade Técnica de Lisboa, July 2001.

[DSA09]   R. Dinis, P. Silva, and T. Araújo. Turbo equalization with cancelation of nonlinear distortion for CP-assisted and zero-padded MC-CDM schemes. *IEEE Trans. Commun.*, 57(8):2185–2189, August 2009.

[DTV00]   D. Dardari, V. Tralli, and A. Vaccari. A theoretical characterization of nonlinear distortion effects in OFDM systems. *IEEE Trans. Commun.*, 48(10):1755–1764, October 2000.

[DW01]   N. Dinur and D. Wulich. Peak-to-average power ratio in high-order OFDM. *IEEE Trans. Commun.*, 49, June 2001.

[ETS06]   ETSI Standard: EN 300 401 V1.4.1. *Radio Broadcasting Systems; Digital Audio Broadcasting (DAB) to mobile, portable and fixed receivers*, January 2006.

[ETS09]   ETSI Standard: EN 300 744 V1.6.1. *Digital Video Broadcasting (DVB); Framing structure, channel coding and modulation for digital terrestrial television*, January 2009.

[FH96]   R. F. H. Fischer and J. B. Huber. A new loading algorithm for discrete multitone transmission. In *Proc. IEEE Global Telecommunications Conference (GLOBECOM'96)*, volume 1, pages 724–728, London, UK, November 1996.

[Gal68]   R. G. Gallager. *Information Theory and Realiable Communication*. Wiley, 1968.

[Goo63]   N. R. Goodman. Statistical analysis based on a certain multivariate complex Gaussian distribution (an introduction). *The Annals of Mathematical Statistics*, 34(1):152–177, March 1963.

[GR07]   I. S. Gradshteyn and I. M. Ryzhik. *Table of Integrals, Series, and Products*. Elsevier Academic Press, 2007.

[Gra90]   R. M. Gray. Quantization noise spectra. *IEEE Trans. Inform. Theory*, 36(6):1220–1244, November 1990.

[GV94]   R. Gross and D. Veeneman. SNR and spectral properties for a clipped DMT ADSL signal. In *Proc. IEEE International Conference on Communications 1994 (ICC'94)*, volume 2, pages 843–847, May 1994.

[HH87]   D. Hughes-Hartog. *Ensemble Modem Structure for Imperfect Transmission Media*. U.S. Patent 4679227, July 1987.

[HHS86]   B. Hirosaki, S. Hasegawa, and A. Sabato. Advanced group-band data modem using orthogonally multiplexed QAM technique. *IEEE Trans. Commun.*, 34(6):587–592, June 1986.

[HP97]   S. Hara and R. Prasad. Overview of multicarrier CDMA. *IEEE Commun. Mag.*, 35(12):126–133, December 1997.

[IEE]   IEEE 802.22 Working Group on Wireless Regional Area Networks. *http://www.ieee802.org/22/*.

[IEE04]   IEEE 802.16-2004. *IEEE Standard for Local and Metropolitan Area Networks – Part 16: Air Interface for Fixed Broadband Wireless Access Systems*, October 2004.

[IEE06]   IEEE 802.16-2005. *IEEE Standard for Local and Metropolitan Area Networks – Part 16: Air Interface for Fixed and Mobile Broadband Wireless Access Systems Amendment 2: Physical and Medium Access Control Layers for Combined Fixed and Mobile Operation in Licensed Bands and Corrigendum 1*, February 2006.

[IS73]   E. Imboldi and G. R. Stette. AM-to-PM conversion and intermodulation in nonlinear devices. *Proceedings of the IEEE*, 61(6):796–797, June 1973.

[JBC95]   K. S. Jacobsen, J. A. C. Bingham, and J. M. Cioffi. Synchronized DMT for multipoint-to-point communications on HFC networks. In *Proc. IEEE Global Telecommunications Conference (GLOBECOM'95)*, volume 2, pages 963–966, November 1995.

[JN84]   N. Jayant and P. Noll. *Digital Coding of Waveforms: Principles and Applications to Speech and Video*. Prentice-Hall, New Jersey, 1984.

[JW96]   A. Jones and T. Wilkinson. Combined coding for error control and increased robustness to system nonlinearities in OFDM. In *Proc. IEEE 46th Vehicular Technology Conference (VTC'96)*, Atlanta, USA, May 1996.

[KB80]   W. E. Keasler and D. L. Bitzer. *High-speed modem suitable for operation with a switched network*. U.S. patent 4,206,320, June 1980.

[KR02]   I. Koffman and V. Roman. Broadband wireless access solutions based on OFDM access in IEEE 802.16. *IEEE Commun. Mag.*, 40(4):96–103, April 2002.

[KRJ00]   B. S. Krongold, K. Ramchandran, and D. L. Jones. Computationally efficient optimal power allocation algorithms for multicarrier communication systems. *IEEE Trans. Commun.*, 48(1):23–27, January 2000.

[LC97]   A. Leke and J. M. Cioffi. A maximum rate loading algorithm for discrete multitone modulation systems. In *Proc. IEEE Global Telecommunications Conference 1997 (GLOBECOM'97)*, volume 3, pages 1514–1518, Phoenix, Arizona, USA, November 1997.

[LC98]   X. Li and L. J. Cimini, Jr. Effects of clipping and filtering on the performance of OFDM. *IEEE Commun. Lett.*, 2(5):131–133, May 1998.

[LSC07]   J. Lee, R. V. Sonalkar, and J. M. Cioffi. Multiuser bit loading for multicarrier systems. *IEEE Trans. Commun.*, 54(7):1170–1174, July 2007.

[LZB08]   J. Li, X. Zhang, and N. Beaulieu. Precise calculation of the SEP of 128- and 512-cross-QAM in AWGN. *IEEE Commun. Lett.*, 12(1):1–3, January 2008.

[MBFH97]   S. Müller, R. Bräuml, R. Fischer, and J. Huber. OFDM with reduced peak-to-average power ratio by multiple signal representation. *Annales of Telecommunications*, 52, February 1997.

[MCDG00]   B. Muquet, M. Courville, P. Dunamel, and G. Giannakis. OFDM with trailing zeros versus OFDM with cyclic prefix: links, comparisons and application to the HiperLAN/2 system. In *Proc. IEEE International Conference on Communications 2000 (ICC'00)*, volume 2, pages 1049–1053, New Orleans, Louisiana, USA, June 2000.

[MCG+00]   B. Muquet, M. Courville, G. B. Giannakis, Z. Wang, and P. Duhamel. Reduced complexity equalizers for zero-padded OFDM transmissions. In *Proc. IEEE International Conference on Acoustics, Speech, and Signal Processing (ICASSP'00)*, volume 5, pages 2973–2976, Istanbul, Turkey, June 2000.

[MH97]   S. Müller and J. Huber. A comparison of peak reduction schemes for OFDM. In *Proc. IEEE Global Telecommunications Conference 1997 (GLOBECOM'97)*, Phoenix, Arizona, USA, May 1997.

[Mit95]   J. Mitola. The software radio architecture. *IEEE Commun. Mag.*, 33(5):26–38, May 1995.

[MKP07]   M. Morelli, C.-C. J. Kuo, and M.-O. Pun. Synchronization techniques for orthogonal frequency division multiple access (OFDMA): A tutorial review. *Proc. IEEE*, 95(7):1394–1427, July 2007.

[MR98]   T. May and H. Rohling. Reducing the peak-to-average power ratio in OFDM radio transmission systems. In *Proc. IEEE 48th Vehicular Technology Conference (VTC'98)*, Ottawa, Canada, May 1998.

[OI00]   H. Ochiai and H. Imai. Performance of deliberate clipping with adaptive symbol selection for strictly band-limited OFDM systems. *IEEE J. Select. Areas Commun.*, 18(11), November 2000.

[OI01]   H. Ochiai and H. Imai. Performance analysis of deliberately clipped OFDM signals. *IEEE Trans. Commun.*, 50(1), January 2001.

[OL95]   R. O'Neill and L. Lopes. Envelope variations and spectral splatter in clipped multicarrier signals. In *Proc. 6th IEEE International Symposium on Personal, Indoor and Mobile Radio Communications (PIMRC'95)*, September 1995.

[Pap84]   A. Papoulis. *Probability, Random Variables and Stocastic Processes*. McGraw-Hill, 1984.

[PBM86]   A. P. Prudnikov, Y. A. Brychkov, and O. I. Marichev. *Integrals and Series, Volume 2 – Special Functions*. Gordon and Breach Science Publishers, New York, NY, 1986.

[Pri58]   R. Price. A useful theorem for nonlinear devices having Gaussian inputs. *IRE Transactions on Information Theory*, 4(2):69–72, June 1958.

[Rap91]   C. Rapp. Effects of HPA-nonlinearity on a 4-DPSK/OFDM-signal for a digital sound broadcasting system. In *Proc. 2nd European Conference on Satellite Communications*, pages 179–184, October 1991.

[Ric45]   S. O. Rice. Mathematical analysis of random noise. *Bell Systems Tech. J.*, pages 46–156, 1945.

[Row82]   H. Rowe. Memoryless nonlinearities with Gaussian inputs: Elementary results. *Bell System Tech. Journal*, 61(7):1519–1525, September 1982.

[Sal67]   B. R. Saltzberg. Performance of an efficient parallel data transmission system. *IEEE Trans. Commun.*, 15:811–815, December 1967.

[SCD+10]   M. M. Silva, A. Correia, R. Dinis, N. Souto, and J. C. Silva. *Transmission Techniques for Emergent Multicast and Broadcast Systems*. CRC Press, 2010.

[Shi71]   O. Shimbo. Effects of intermodulation, AM-PM conversion, and additive noise in multicarrier TWT systems. *Proc. IEEE*, 59(2):230–238, February 1971.

[SLK97]   H. Sari, Y. Levy, and G. Karam. An analysis of orthogonal frequency-division multiple access. In *Proc. IEEE Global Telecommunications Conference 1997 (GLOBECOM'97)*, Phoenix, Arizona, USA, November 1997.

[Ste74]   G. Stette. Calculation of intermodulation from a single carrier amplitude characteristic. *IEEE Trans. Commun.*, 22(3):319–323, March 1974.

[Sze75]   G. Szego. *Orthogonal Polynomials*, volume 23 of *The Art of Computer Programming*. Colloquium Publications, 4th edition, 1975.

[TBT99]   T. Turletti, H.J. Bentzen, and D. Tennenhouse. Toward the software realization of a gsm base station. *IEEE Journal on Selected Areas in Communications*, 17(4):603–612, April 1999.

[THM72]   D. Tufts, H. Hersey, and W. Mosier. Effects of FFT coefficient quantization on bin frequency response. *Proc. IEEE*, 60(1):146–147, January 1972.

[vdB95]   A. van den Bos. The multivariate complex normal distribution – a generalization. *IEEE Trans. Inform. Theory*, 41(2):537–539, March 1995.

[vNP00]   R. van Nee and R. Prasad. *OFDM for Wireless Multimedia Communications*. Artech House Publications, 2000.

[Wal99]   R. H. Walden. Analog-to-digital converter survey and analysis. *IEEE Journal on Selected Areas in Communications*, 17(4):539–550, April 1999.

[WE71]   S. B. Weinstein and P. M. Ebert. Data transmission by frequency-division multiplexing using the discrete Fourier transform. *IEEE Trans. Commun. Technol.*, 19(5):628–634, October 1971.

[WKL96]   B. Widrow, I. Kollar, and M.-C. Liu. Statistical theory of quantization. *IEEE Trans. Instrum. Meas.*, 45(2):353–361, April 1996.

[YLF93]   N. Yee, J. P. Linnartz, and G. Fettweis. Multi-carrier CDMA in indoor wireless radio networks. In *Proc. 4th International Symposium on Personal, Indoor and Mobile Radio Communications (PIMRC'93)*, pages 109–113, Yokohama, Japan, September 1993.

[ZK99]   K. C. Zangi and R. D. Koilpillai. Software radio issues in cellular base stations. *IEEE Journal on Selected Areas in Communications*, 17(4):561–573, April 1999.

[Zwi03]   D. Zwillinger. *CRC Standard Mathematical Tables and Formulae*. Chapman & Hall/CRC, Boca Raton, Florida, 2003.

Index